I0885332

Utopian and Dystopian Explorations of Pandemics and Ecological Breakdown

This edited collection, which is situated within the environmental humanities and environmental social sciences, brings together utopian and dystopian representations of pandemics from across literature, the arts, and social movements.

Featuring analyses of literary works, TV and film, theater, politics, and activism, the chapters in this volume home in on critical topics such as posthumanism, multispecies futures, agency, political ecology, environmental justice, and Indigenous and settler-colonial environmental relations. The book asks: how do pandemics and ecological breakdown show us the ways that humans are deeply interconnected with the more-than-human world? And what might we learn from exploring those entanglements, both within creative works and in lived reality? Brazilian, Indian, Polish, and Dutch texts feature alongside classic literary works like Defoe's *A Journal of a Plague Year* (1722) and Matheson's *I Am Legend* (1954), as well as broader takes on movements like global youth climate activism. These investigations are united by their thematic interests in the future of human and nonhuman relationships in the shadow of climate emergency and increasing pandemic risk, as well as in the glimmers of utopian hope they exhibit for the creation of more just futures.

This exploration of how pandemics illuminate the entangled materialities and shared vulnerabilities of all living things is an engaging and timely analysis that will appeal to environmentally minded researchers, academics, and students across various disciplines within the humanities and social sciences.

Heather Alberro is a Senior Lecturer in Global Sustainable Development in the Department of History, Heritage and Global Cultures, Nottingham Trent University. She also serves as co-convenor for the Political Studies Association's (PSA) environmental politics specialist group and as a member of the PSA's Executive Committee.

Emrah Atasoy, an Associate Professor of English, is a Marie Skłodowska-Curie Fellow (EUTOPIA-SIF COFUND) of the Institute of Advanced Study (IAS), working in the Department of English and Comparative Literary Studies, University of Warwick, UK.

Nora Castle is currently a Lecturer at the University of Bonn. She received her PhD in 2023 from the Department of English and Comparative Literary Studies, University of Warwick, UK, where she also completed an Early Career Fellowship at the Institute of Advanced Study.

Rhiannon Firth is a Lecturer in Sociology of Education at IoE, UCL's Faculty of Education and Society. She co-leads the MA modules Sociology of Education and Gender, Sexuality and Education, and is Program Leader for the MA in Sociology of Education.

Conrad Scott, PhD, is an Associate Lecturer in the Department of English and Film Studies, University of Alberta, and is an Individualized Study Tutor for the University of Athabasca's Honours English course "The Ecological Imagination," where he holds an SSHRC Postdoctoral Fellowship.

"Fully globalized, immediately connected, yet still radically unequal in resources and protections, humanity has now become aware of itself as a species in a new, more urgent way: when pathogens and environmental disruptions strike, how can past experiences and their representations provide perspective, balance, and hope? This book provides answers."

James Engell, *Gurney Professor of English and Professor of Comparative Literature, Harvard University, USA*

"This book is a collection of diverse and passionately engaged explorations of the way we live now. It is imbued both with a sense of the traumas of (post) apocalypse and a hope that human and non-human species can find ways to survive into futures that are not simply continuations of a present scarred by pandemics, extinctions, and the eco-injustices of global capital. The essays here are international in scope, multiplex in their critical methodologies, and comprehensive in their coverage. They provide resources for thinking about how to move into futures in which, through this "breakdown," we take our non-anthropocentric place as one of the many species co-existing in an ecosystem that encompasses all life on the planet."

Veronica Hollinger, *Editor, Science Fiction Studies, and Professor Emerita of Cultural Studies, Trent University, Canada*

"As the introduction describes, this timely book emerged out of a dark and precarious contemporary moment, in the world and in the field of utopian studies. By bringing together this collection of cutting-edge studies by such a diverse mix of scholars addressing one of the most disruptive and destructive events of recent times, the editors have delivered an insightful and impactful counterpoint to official and normative invitations to despair and capitulate. This volume is itself an act of utopian annunciation in the face of official denunciation. Read it, hope, and act."

Tom Moylan, *Professor Emeritus in the School of English, Irish, and Communication, and member of the Ralahine Centre for Utopian Studies, University of Limerick, Ireland*

Routledge Environmental Humanities

Series editors: Scott Slovic (University of Idaho, USA), **Joni Adamson** (Arizona State University, USA) and **Yuki Masami** (Aoyama Gakuin University, Japan)

The *Routledge Environmental Humanities* series is an original and inspiring venture recognising that today's world agricultural and water crises, ocean pollution and resource depletion, global warming from greenhouse gases, urban sprawl, overpopulation, food insecurity and environmental justice are all *crises of culture*.

The reality of understanding and finding adaptive solutions to our present and future environmental challenges has shifted the epicenter of environmental studies away from an exclusively scientific and technological framework to one that depends on the human-focused disciplines and ideas of the humanities and allied social sciences.

We thus welcome book proposals from all humanities and social sciences disciplines for an inclusive and interdisciplinary series. We favour manuscripts aimed at an international readership and written in a lively and accessible style. The readership comprises scholars and students from the humanities and social sciences and thoughtful readers concerned about the human dimensions of environmental change.

For more information about this series, please visit: www.routledge.com/Routledge-Environmental-Humanities/book-series/REH

Utopian and Dystopian Explorations of Pandemics and Ecological Breakdown

Entangled Futurities

Edited by
Heather Alberro, Emrah Atasoy, Nora Castle, Rhiannon Firth, and Conrad Scott

Routledge
Taylor & Francis Group
LONDON AND NEW YORK

earthscan
from Routledge

First published 2025
by Routledge
4 Park Square, Milton Park, Abingdon, Oxon OX14 4RN

and by Routledge
605 Third Avenue, New York, NY 10158

Routledge is an imprint of the Taylor & Francis Group, an informa business

Funding was received from University College London.

British Library Cataloguing-in-Publication Data
A catalogue record for this book is available from the British Library

Library of Congress Cataloging-in-Publication Data
Names: Alberro, Heather, editor. | Atasoy, Emrah, editor. |
Castle, Nora, editor. | Firth, Rhiannon, editor. | Scott, Conrad, editor.
Title: Utopian and dystopian explorations of pandemics and ecological breakdown : entangled futurities / edited by Heather Alberro, Emrah Atasoy, Nora Castle, Rhiannon Firth, and Conrad Scott.
Description: London ; New York : Routledge, 2024. | Series: Routledge environmental humanities | Includes bibliographical references and index.
Identifiers: LCCN 2024016720 (print) | LCCN 2024016721 (ebook) |
ISBN 9781032385914 (hardback) | ISBN 9781032385938 (paperback) |
ISBN 9781003345770 (ebook)
Subjects: LCSH: Pandemics in the performing arts. | Pandemics in literature. | Pandemics—Social aspects. | Human ecology. | Ecocriticism.
Classification: LCC PN1590.P36 U86 2024 (print) |
LCC PN1590.P36 (ebook)| DDC 701.3—dc23/eng/20240417
LC record available at https://lccn.loc.gov/2024016720
LC ebook record available at https://lccn.loc.gov/2024016721

ISBN: 978-1-032-38591-4 (hbk)
ISBN: 978-1-032-38593-8 (pbk)
ISBN: 978-1-003-34577-0 (ebk)

DOI: 10.4324/9781003345770

Typeset in Times New Roman
by codeMantra

Contents

Editor Bios

Heather Alberro is a Senior Lecturer in Global Sustainable Development in the Department of History, Heritage and Global Cultures, Nottingham Trent University. Her background and interests span a range of disciplines including green utopianism, critical posthuman theory, post-anthropocentric and normative environmental ethics, literary ecocriticism, and radical environmental politics. Her publications include "In and Against Eco-Apocalypse: On the Terrestrial Ecotopianism of Radical Environmental Activists" in the journal *Utopian Studies*, "H.G Wells, Earthly and Post-Terrestrial Futurities" (2022) in the journal *Futures*, and the monograph *Terrestrial Ecotopias: Multispecies Flourishing in and Beyond the Capitalocene* (Peter Lang, 2024). Heather also serves as co-convenor for the Political Studies Association's (PSA) environmental politics specialist group, and as a trustee member of the PSA's executive board.

Emrah Atasoy, an Associate Professor of English, is a Marie Skłodowska-Curie Fellow (EUTOPIA-SIF COFUND) of the Institute of Advanced Study (IAS), working in the Department of English and Comparative Literary Studies, University of Warwick, UK. He was a visiting researcher at Harvard University's Department of English, US from February to May 2024, the University of Pompeu Fabra in Barcelona, Spain from September to November 2023, and the University of Oxford's Faculty of English Language and Literature from September 2021 to September 2022. He is the author of the monograph *Epistemological Warfare and Hope in Critical Dystopia* (Nobel, 2021). His work appeared in journals such as *Critique: Studies in Contemporary Fiction* (with Marta Komsta), *Studies in the Novel* (with Thomas Horan), *Utopian Studies*, *Librosdelacorte.es*, *Literary Voice*, *Methis. Studia Humaniora Estonica*, *SFRA Review*, and *Journal of the Fantastic in the Arts*. He also contributed chapters to *The Postworld In-Between Utopia and Dystopia: Intersectional, Feminist, and Non-Binary Approaches in 21st-Century Speculative Literature and Culture* (Routledge, 2021) and *Speculations of War: Essays on Conflict in Science Fiction, Fantasy and Utopian Literature* (McFarland, 2021). He recently co-edited (together with Burcu Kayışcı Akkoyun and Merve Tabur) the critical forum "Cultural Encounters and Textual Speculations in the Mediterranean," published in the journal *Utopian Studies* (vol. 35:1, 2024).

Nora Castle is currently a Lecturer at the University of Bonn. She received her PhD in 2023 from the Department of English and Comparative Literary Studies, University of Warwick, UK, where she also completed an Early Career Fellowship at the Institute of Advanced Study. Her research interests include science fiction studies, food futures, critical animal studies, critical plant studies, and critical infrastructure studies. She is the coeditor of an edited collection on *Animals and Science Fiction* (Palgrave, 2024), as well as a special issue on "Food Futures" in *Science Fiction Studies* (vol. 49:2, 2022) and a special issue on "Animal Futurity" in *Green Letters* (vol. 61:1, 2022), among other projects. Her recent publications include the chapter "Nonhuman Entanglements in Adam Roberts' Science Fiction: *Bête* (2014) and *By Light Alone* (2012)" in *Literature and Politics of the Environment* (Boydell & Brewer, 2023) and the open access article "In Vitro Meat and Science Fiction: Contemporary Narratives of Cultured Flesh" in *Extrapolation* (vol. 63:2, 2022), the latter of which won an Honorable Mention for the 2023 SFRA Innovative Research Award.

Rhiannon Firth is a Lecturer in Sociology of Education at IoE, UCL's Faculty of Education and Society. She coleads the MA modules Sociology of Education and Gender, Sexuality and Education, and is Program Leader for the MA Sociology of Education. Rhiannon's research is at the intersection of sociology, education, and politics with a focus on the pedagogical and prefigurative practices of social and ecological justice movements, and on grassroots organizing within, against and beyond capitalist crises and disasters. She has conducted funded research with self-managed sustainable communities including co-operatives and eco-villages; with grassroots disaster relief movements including Occupy Sandy and COVID-19 Mutual Aid UK; and with alternative organizations using automation in experimental and artisanal ways including a co-operative and a hackspace. She has also conducted historical and theoretical research on feminist consciousness-raising, utopian literature and communities, critical pedagogy, and critical cartography. She is the author of three previous books: *Utopian Politics: Citizenship and Practice* (Routledge, 2012), *Coronavirus, Class and Mutual Aid in the UK* (Palgrave, 2020), and *Disaster Anarchy: Mutual Aid and Radical Action* (Pluto, 2022) alongside several journal articles in outlets including *ephemera*, *Gender and Education*, and *Time and Society*, as well as book chapters and shorter pieces in media and activist outlets. Before coming to IoE, Rhiannon worked in research and teaching roles at the University of Essex, University of East London, and Brunel University. She completed her ESRC-funded PhD at the University of Nottingham in 2010.

Conrad Scott, PhD, is an Associate Lecturer in the Department of English and Film Studies, University of Alberta and is an Individualized Study Tutor for the University of Athabasca's Honours English course on "The Ecological Imagination," where he holds an SSHRC Postdoctoral Fellowship. He researches contemporary speculative fiction and environmental literature, with current projects focused on plant and animal futures, as well as the spatial turn. His reviews and essays have appeared in *Science Fiction Studies*, *Extrapolation*,

Paradoxa, *Transmotion*, *Environmental Philosophy*, *The Goose*, *UnderCurrents*, *Canadian Literature*, *The Routledge Handbook of CoFuturisms* (2023), and *Animals and Science Fiction* (Palgrave, 2024). He is also the author of *Waterline Immersion* (Frontenac House, 2019), and serves as the Science Fiction Research Association's Country Rep for Canada.

Contributor Bios

Sara González Bernárdez (she/her) is a graduate in English Language and Literature from the University of Santiago de Compostela, and holds a master's and a PhD in Advanced English Studies from the same institution. Her thesis deals with the possibilities which the fantasy genre affords as a site of resistance and expression for marginalized identities, using the work of N. K. Jemisin as a case study. Her research thus revolves around the concepts of identity and representation in fiction, examined from the particular lens of feminist criticism. Though she works mostly with contemporary fantasy literature, her interests also include new forms of media, such as videogames, and fandom studies.

Marleen Boschen holds a PhD in Cultural Studies from Goldsmiths, University of London. She is a Lecturer at Chelsea College of Art and Design with a focus on art and ecology. Her doctoral research looked at questions of care and conflict in seed banking practices across the globe. Marleen is also a practicing artist and curator, and has coedited the collections *Agropoetics Reader* (SAVVY Contemporary & TIER, 2020) and *Seeds* (Biennale Gherdëina, 2022).

Alice Breemen is a theater scholar and philosopher based in Amsterdam, The Netherlands, where she is writing her dissertation about dramaturgies of the Anthropocene. The relational entanglements of art, climate, and future are central in her analyses of contemporary theatrical performances. She did research at the Freie Universität Berlin with a DAAD Scholarship and organized an alternative online study trip themed Theatre and the Anthropocene for the UvA Bachelor Theatre Studies students. In 2023, she received the UvA Finishing Fellowship for completing her thesis.

Benjamin Burt is an independent researcher whose work draws from the fields of utopian studies, ecocriticism, and urban studies to examine cultural attitudes toward contemporary urban life and environmental transformation in Brazil, Latin America, and Lusophone Africa. He earned his PhD in Hispanic Languages and Literatures from UCLA in 2020 and subsequently held roles as Visiting Assistant Professor at Pitzer College, Research Fellow at the UCLA Center for Brazilian Studies, and adjunct faculty at UCLA. His most recent publications can be found in *Journal of Lusophone Studies, Fragmentum*, and *Brasil/Brazil*.

Tânia Cerqueira is currently a PhD candidate at the Faculty of Arts and Humanities of the University of Porto (Portugal). Her thesis, funded by National Funds through FCT—Foundation for Science and Technology (2021.04547.BD), explores the relationship between the Gothic tradition and English fictional young adult dystopias. She collaborates at the Centre for English, Translation, and Anglo-Portuguese Studies (CETAPS) and is an Executive Board member of the Young Adult Studies Association (YASA). Her main research interests include young adult fiction, dystopian narratives, Gothic literature and culture, and posthumanism.

Claire P. Curtis is Professor of Political Science at the College of Charleston. She is the author of *Postapocalyptic Fiction and the Social Contract: "We'll Not Go Home Again"* (Lexington, 2010) and the coeditor of the special issue of *Utopian Studies* dedicated to Octavia Butler (19.3, 2008). She has published articles on Ursula Le Guin, Octavia Butler, and contemporary young adult postapocalyptic fiction. Most recently, she published "Standards of Justice for Human Being and Doing in Kim Stanley Robinson's *2312* and C. S. Friedman's *This Alien Shore*," *Open Library of Humanities*, 3(2): 6 (2017) and "The Politics of Living Together: Butler's Short Stories and Teaching Political Philosophy" in *Approaches to Teaching the Works of Octavia E. Butler* (Modern Language Association, 2019). Currently, she is writing a book on Martha Nussbaum's capabilities approach and contemporary postapocalyptic fiction. As a political philosopher, she uses fiction as the experimental space of living together to analyze theories of justice. She teaches courses in the history of political thought; utopia/dystopia; and the contemporary intersections of gender, theory, and law. She lives in Charleston, SC.

Rashmi Gaur is a distinguished Professor of English at the Department of Humanities and Social Sciences, Indian Institute of Technology Roorkee, India. She teaches courses on gender and new media studies, body language, and the ethical interfaces of AI and society from a posthumanist perspective. She has created four NPTEL (National Programme on Technology Enhanced Learning) courses, including Contextualizing Gender and Communication in the Digital Age. She has published her work in reputed journals and varied academic forums. At present, she is working on the processes and discourses on higher education pedagogy and practices in India.

Hanna Grześkiewicz is an independent researcher, curator, and activist. Her research focuses on activism, sound, feminism, and border(lessnes)s. She has degrees from the University of Cambridge and the Humboldt University in Berlin, and teaches at the University of the Arts in Berlin.

Benjamin Horn is a self-funded PhD student at the University of Birmingham, UK. His thesis concerns the science fiction of Philip K. Dick, and speculative realist literary theory, to demonstrate how Dick's fiction instantiates a nonhuman real, distinguished from the spurious dialectic of the textual and the material in postmodern capitalism, thereby augmenting readers' relationship to historical novelty. Ben has presented papers at the Productive Futures conference (2018) and the "Living in the End Times" conference (2019). He has cowritten for Routledge with Jayde Martin in the edited collection *Posthuman Pathogenesis* (2022), and has reviews published in *Fantastika*, *SFRA*, and a forthcoming one for *Fafnir*. Ben has taught undergraduates and postgraduates at the University of Birmingham, and as part of extra-academic seminar groups. He is a co-founder of the Contemporary Theoretical and Literature Network (CTRL Net) at the University of Birmingham. Ben's research specialisms are science fiction, posthumanism, digital humanities, speculative realism, and contemporary Marxism.

Rylan Kafara (he/him/his) is a PhD Candidate at the University of Alberta in the Faculty of Kinesiology, Sport, and Recreation. He holds a BA and MA in history, and is a community-based, activist ethnographer. His current research focuses on unsanctioned encampments and how forms of creativity, mutual care, and community support are used by encampment residents to both navigate the challenges of houselessness and enrich their lives.

Jari Käkelä, PhD, has focused his research extensively on the work of Isaac Asimov, other Golden age authors, themes of frontier, and pulp publishing. His current research interests include works of Sarah Pinsker and Becky Chambers, and examines the intellectual linkage between, and challenges by, twenty-first-century hopepunk and Golden Age techno-optimism. He currently works as a Lecturer at the University of Helsinki, is the reviews editor for *Fafnir—Nordic Journal of Science Fiction and Fantasy Research*, and has published in *Extrapolation* and *Journal of the Fantastic in the Arts*, among others.

Ujjwal Khobra is a Doctoral Research Scholar at the Department of Humanities and Social Sciences, Indian Institute of Technology Roorkee, India. Her research work broadly focuses on reading posthuman representations in select Indian Speculative Fiction in English, and posthuman entanglements in everyday life post the COVID-19 pandemic. She has presented research papers on pandemic gardening and phobic responses to contagiosity in many international conferences and academic forums. She has published her research paper entitled "Live (Life) Streaming: Virtual Interaction, Virtual Proximity, and Streaming."

Matthew Leggatt is a Senior Lecturer in English and American Literature at the University of Winchester, UK. He is the author of *Cultural and Political Nostalgia in the Age of Terror* (Routledge, 2018) and the editor of the collection *Was It Yesterday? Nostalgia in Contemporary Film and Television* (SUNY Press, 2021). He has also authored several articles and book chapters including "'Another World Just Out of Sight': Remembering or Imagining Utopia in Emily St John Mandel's *Station Eleven*" (*Open Library of Humanities*, 2018) and "'We Delivered the Bomb': On Jaws, Guilt, and the Atomic Myth" in *The Jaws Book* (Bloomsbury, 2020). He is currently editing the collection *Wastelands and Wonderlands: Essays on Utopian and Dystopian Film and Literature* (SUNY Press, 2024).

Jayde Martin is a PhD student at the University of Birmingham, funded by the Arts and Humanities Research council (UKRI). Her thesis focuses on genetic manipulation, neo-eugenics, and identity in feminist science fiction by Margaret Atwood, Octavia Butler, and Nancy Kress. She uses posthumanist critical theory to understand the reception of the genome as both a concept and a metaphor in literature and culture, versus its scientific reality. Jayde's work analyzes the interactions between science fiction texts and discourses of science and science communication, contemporaneous with these authors' publications, in an interdisciplinary manner. As an academic exploring and analyzing popular culture, she founded the Midlands Network of Popular Culture (2020–2022) and the Gothica Reading Group and Lecture Series (2019–2021). Jayde has also cowritten on science fiction texts with Ben Horn for Routledge's edited collection *Posthuman Pathogenesis* (2022) and has copresented with him for the Living in the End Times conference (2019). These papers explore embodiment, entanglement, social theory, and evolution in literature.

Heather McKnight has a PhD in Law Studies from the University of Sussex and is currently a research consultant for the Magnetic Ideals Collective. She is a critical utopian scholar and activist with research interests in activism, unions, school strikes, education, race, feminism, gender, and speculative fiction.

Timothy S. Murphy (he/him), Houston-Truax-Wentz Professor and Regents Professor of English at Oklahoma State University, is the author of *Wising Up the Marks: The Amodern William Burroughs* (1997), *Antonio Negri: Modernity and the Multitude* (2012), and *William Hope Hodgson and the Rise of the Weird: Possibilities of the Dark* (2023), as well as numerous articles, book chapters, and reviews on modern and contemporary literature, science fiction and fantasy, and critical theory. He has translated works of philosophy and critical theory by Antonio Negri, Gilles Deleuze, Jacques Derrida, Luisa Muraro, Massimo Cacciari, Franco "Bifo" Berardi, and others from French and Italian, and from 2000 to 2013, he served as the general editor of the scholarly journal *Genre: Forms of Discourse and Culture*.

Eleonora Rossi recently defended her PhD in English and Humanities, titled "Becoming Water: Posthuman Feminism in Contemporary Speculative Fiction," at the University of London. Her current research focuses on instances of domestic and urban infiltration which expose and challenge logics and practices of anthropocentric (as well as colonial and heteropatriarchal) ordering of space. This new direction builds on her MA in Sociology (London), during which she explored the gendering of urbanity, through the lens of art and literature. As an unexceptional but avid city runner, Dr. Rossi is endlessly fascinated by the sets of ongoing (and often more-than-human) negotiations that are required to craft and protect human lived spaces—particularly as we navigate the uncertain waters of accelerating climate change. You can currently find her plodding along the banks of the Spree and the Rhein, thinking hard thoughts about water pollution, river-based communities, and unruly sweet water tides.

Alice Seville is a critical PhD student with EDACS at the University of Birmingham, UK. Her thesis concerns post-millennial poetry and ecocritical poetry criticism, plus the interplay of virtual aesthetics and material aesthetics within a wider thematic of cyberpunk literature. Alice is applying these terms and ideas to the ecocritical and posthuman poetics of contemporary English language poets such as Stephen Sexton and Patricia Lockwood, among others. Alice recently interviewed Stephen Sexton for *Spit the Pips*, the Irish literature journal. Alice's own poetic work, which favors formal experimentation, cut-up, and post internet abstraction of emotional registers and meaning, can be found in *One Hand Clapping* magazine. Alice also founded the network PGR Feminisms, which ran from 2019 to 2022 and fostered discussion groups among postgraduates concerning prescient topics for contemporary feminist discourse (in both the pop cultural and academic sectors or uniting the two).

Hülya Yağcıoğlu is an Associate Professor in the College of Humanities and Social Sciences, Zayed University, Abu Dhabi, UAE. She earned her PhD degree in English Literature from Boğaziçi University, Turkey, in 2015. She has published book chapters and articles in the field of comparative literature, specifically on the interaction between material culture and literature. Her research areas include comparative literature, thing theory and material culture in literature, and object-oriented ontology. Her most recent publications include the book chapter "Reifying Innocence: Material Contexts of Love in The Age of Innocence and The Museum of Innocence" in *The Materiality of Love: Essays on Affection and Cultural Practice* (Routledge, 2018) and the coedited book *All Things Arabia: Arabian Identity and Material Culture* (Brill, 2020).

Introduction

Entangled Futurities

Heather Alberro, Emrah Atasoy, Nora Castle, Rhiannon Firth, and Conrad Scott

In this current moment that scientists have dubbed the "Anthropocene"—a term often critiqued for its nomenclature, with even the Chair of the Anthropocene Working Group, Jan Zalasiewicz, noting that "there are now many Anthropocenes out there" (124)—it is clear that humanity has indeed altered the stratigraphy of the planet (Szerszynski 169–70; Zalasiewicz et al., "Stratigraphy"). This process has, in turn, perpetuated what Naomi Klein names "sacrifice zones," which "purport to disentangle the web of life" (Farrier 53). But even as we dissolve the ostensible boundary between humanity and the natural world, where, as David Quammen argues, "environmental disruption by humans [is] a releaser of epidemics" (*Spillover* 237), we are continually reminded that ideas like "nature" (Morton) and "wilderness" (Cronon; see also Hulme 40) are misnomers: humans are not separate from the web of life, and wilderness is not really "wild." Building on Jason Moore's understandings of "intimacy, porosity, and permeability" (*Capitalism* 7), Karen Barad's concept of "intra-action," and Donna Haraway's inclusion of "myriad intra-active entities-in-assemblages[,] including the more-than-human, other-than-human, inhuman, and human-as-humus" ("Anthropocene" 160), *Utopian and Dystopian Explorations of Pandemics and Ecological Breakdown: Entangled Futurities* explores the breakdown of the human/nonhuman binary through the figures of pandemics, epidemics, and other virulent health crises across the utopian and dystopian spectrum.

The volume is purposefully interdisciplinary and varied, with research stemming from both the humanities and social sciences and with authors hailing from a variety of different theoretical schools and backgrounds. The chapters in this volume home in on readings of historical, current, and future pandemics from a wide range of approaches that, collectively, contemplate our relationships with the natural world, ourselves, and this planet that sustains us. How, the book asks, do pandemics show us the ways that humans are—and have always been—deeply interconnected with the other-than-human world? And what might we learn from exploring those entanglements, both within creative works and in lived reality? Through their work on different kinds and valences of entanglements, the authors of the volume highlight humans' interdependence and connectivity with (the) other-than-human world(s), as well as the various methods of creative, critical, and political action humans have taken to obscure, understand, and/or rewrite

DOI: 10.4324/9781003345770-1

those connections. Moore asserts that “we make environments and they make us” (“Capitalocene” 599). As a whole, the “entangled futurities” of this book underscore the reality that humanity is but part of a functional—and sometimes dysfunctional—whole. By employing the term “futurities,” we underscore the complicated nature of utopianism and dystopianism explored within these chapters; the nature of our future hinges on our actions now, and the fallout could be in either direction. Our health, and our continued existence, depends on creating environments that enable the well-being of all, not merely a privileged few. The volume argues for the importance of acknowledging this reality and, significantly, taking steps to act—politically, culturally, infrastructurally—to create more just futures for all, not only for the sake of human health but also for the continuing health of the planet.

This book has come out of a complicated and difficult time. Its origins began percolating in 2019, when many of the editors met at the Utopian Studies Conference in Prato, Italy. Not long after came the onset of the COVID-19 pandemic. Out of the dark of the pandemic, the light of collaborative work shined through, and the five of us, as well as five others, co-organized the Living in the End Times: Utopian and Dystopian Representations of Pandemics in Fiction, Film, and Culture conference in January 2021 (see Castle et al), hosted online on Teams, Zoom, and YouTube Live, and supported by Cappadocia University, Türkiye. The conference, which featured over 200 speakers from over 40 different countries, was a testament to both the power of community and the power of technology (Zoom bombing notwithstanding) to bring people together in alienating times. As Kim Stanley Robinson noted in his keynote address, the conference was “both a symptom and a diagnostic of the time that we’re in” (“Keynote II”). This gathering sought to explore not only representations of pandemics in a variety of narratives but also to examine what might come after. It was an attempt to search for and nurture that faint glimmer of hope among the despair and destruction, that “insurgent hope of … utopian impulse with its transformative capacity” (Moylan 1). That goal is one that we have taken forward into this volume, which ends, as will be described shortly, with Part 4, Creative Resistance and Utopian Glimmers. The conference led to a special section in the *SFRA Review* (vol. 51, no. 2), which, in turn, led to the development of this volume. The book soon developed a life of its own beyond the conference—as any good book connected to posthuman movements should. Our contributors therefore include not only a number of scholars who attended and presented at the conference, including some who were part of the *SFRA Review* section, but also fresh faces who have added their voices to the dys/utopian chorus. We are incredibly grateful to our authors, including those who joined this vessel further downstream.

While this volume exists largely in response to the COVID-19 pandemic, its development was also stalled by it, with many of us, as well as our contributors, falling ill—often several times. Many of us have also experienced financial precariousness in the seemingly unending search for secure employment that characterizes our dystopian neoliberal hellscape. As we write this in January 2024, the continued timeliness of the volume is striking; we may no longer be “peak COVID,” but while the pandemic has faded somewhat from public consciousness, it has certainly not gone away. Its reverberations are still being, and will continue to be, felt, from the panic induced by a sniffly nose to the increased awareness of the fragility of our just-in-time supply chains and beyond. And with climate change increasing the probability of future pandemics (Marani et al.), it is unlikely that the topic of pandemics will cease to be relevant, both within and outside of academic discussion, any time soon. The vast range of creative and critical works, performances, and political actions responding to pandemics and climate emergency will likewise only continue to proliferate. Even if we had several books worth of space, it would be impossible to do justice to the diversity of works we have already encountered on our journey—let alone the works that are yet to come. We can only imagine (and look forward to) the brilliant work that scholars may do on the topic in the future. The 15 chapters in this collection therefore represent only a fraction of the work that has and could be done, but we hope its diversity will inspire additional scholarly research in a variety of directions. The authors practice their scholarship in institutions from a wide range of countries, including Brazil, Canada, Finland, Germany, India, the Netherlands, Portugal, Spain, United Arab Emirates, the UK, and the US, and their research subjects range from post-humanism to ecocriticism, from monstrosity to ecofeminism, from ecoGothic to ecophobia, from classic zombies and monsters to fungally mediated apocalypse, from environmental justice to creative practice, and from intersectional critique to utopian glimmers. We hope that this edited collection will prompt us all to seek out more utopian perspectives and spaces in, against, and beyond the end times.

Key Concepts

Our collective obsession with pandemics arose out of the historical turn of events we found ourselves within, starting in late 2019/early 2020. As Robinson asserted in 2020 while referencing Raymond Williams' claim that "every historical period has its own 'structure of feeling'" (Williams as qtd. in Robinson, "Coronavirus"), "the coronavirus is rewriting our imaginations" (Robinson, "Coronavirus"). COVID-19 was—and still is—a globally catastrophic event and a proliferation of academic, literary, and other cultural objects arose as products of this time. This includes work like Louise Erdrich's *The Sentence* (2021), Sarah Moss's *The Fell* (2021), Ali Smith's *Companion Piece* (2022), Clare Pollard's *Delphi* (2022), and Isabel Allende's *Violeta* (2022), to name just a few. Nevertheless, COVID-19 was not the first, nor will it be the last pandemic, and, in fact, pandemics have been a concern within literary and religious works that vastly predate it. These go back as far as Thucydides' account of the fifth century BCE "Plague of Athens" (Yeomans) and the Old Testament, with its plagues and foretelling of pestilence in the end of days. Giovanni Boccaccio's *Decameron* (1353), Petrarch's letters (mid-to-late fourteenth century), and Geoffrey Chaucer's *The Canterbury Tales* (late fourteenth century) all actively contend with the aftermath of the Black Death (1346–1353), while later writers of the 1920s, such as Ernest Hemingway and John Dos Passos, avoid mentioning the Great Influenza (1918–1920) in lieu of the Great War, despite the former killing more people than the latter (Hovanec 1)—though the topic of the Spanish Flu was later taken up by writers such as William Maxwell (*They Came Like Swallows,* 1937) and Katherine Anne Porter (*Pale Horse, Pale Rider*, 1939). Later pandemics, such as the Great Plague of London (1665–1666) and the third cholera pandemic (1846–1860), also inspired literature such as Daniel Defoe's *A Journal of the Plague Year* (1722), Mary Shelley's *The Last Man* (1826), and Albert Camus's *The Plague* (1947), respectively—an admittedly Anglocentric list that is but a snapshot of how pandemics have affected literature and cultural expression globally, including through devastating colonial contact. This is unsurprising since pandemics have been an ongoing feature in the story of humanity, and while many of our generation may have dismissed them as a relic relegated to history books by pharmaceutical progress, the events of 2019 proved that assumption wrong. Utopian/dystopian works such as Emily St. John Mandel's *Station Eleven* (2014), Ling Ma's *Severance* (2018), Kevin Chong's *The Plague* (2018), and Sarah Pinsker's *A Song for a New Day* (2019) turned out to be eerily prescient. Indeed, pandemic events look set not only to continue to catch us by surprise, but even to increase drastically as forest destruction by multinationals eliminates the barrier between the human population and viruses endemic to reclusive wild animals; as factory farming incubates novel viruses; as human slums produce "densely packed and immune compromised" populations; and as global inequality increasingly means that the majority of humanity lack access to healthcare and often proper nutrition (Davis 17). Mike Davis argues that "plagues" are endemic to capitalism, and that "permanent bio-protection" against the plagues of capitalism would require wholescale and revolutionary structural reform, since vaccines are not enough (18). In *Embodying Contagion*, Megen de Bruin-Molé and Sara Polak explain that "in today's always-in-crisis culture, outbreak is our shared mode of discourse" (2). Our language is cluttered with contagion analogies, and has even been made the method of contagion in works like the Turkish mini-series *Hot Skull* (original: *Sıcak Kafa*, 2022). This proliferation of analogy reflects the deep hold and influence that pandemics have on human experience, illuminating the ways that humans are *always* entangled with, and contaminated by, one another and the nonhuman world. Pandemics, in particular COVID-19, have heightened our awareness that contact with one another holds both promise and peril; we are leaky, porous bodies, not the self-contained and secure autonomous units described in many Global North cosmologies.

The IPCC sixth assessment report (2023) sounded the alarm over global temperatures now at 1.1C above pre-industrial levels, warning of the "rapidly closing window of opportunity to secure a liveable and sustainable future for all." July 2023 shattered global heat records, and saw apocalyptic floods, wildfires, and droughts ravaging much of the globe (Carrington), prompting UN Secretary General António Guterres to declare the arrival of the era of "global boiling" (Niranjan). Equally, if not more disturbingly, the World Wildlife Fund's "Living Planet" report (2022) documented a near-incomprehensible 69% average decline in monitored wildlife populations since 1970. The risk of pandemics has increased significantly in the wake of land-use changes, wildlife overexploitation, intensive livestock production, and climate change (Lawler et al. 840). Of the novel or re-emerging infectious diseases that have afflicted humans in the twenty-first century, an estimated 75% have been zoonotic in origin (specifically from other vertebrates), while globally, known events of zoonotic disease emerging from wildlife have increased considerably over the past 80 years (Lawler et al. 840; Schmeller et al). Without a radical reorganization of endless-growth-oriented industrial-capitalist societies fueling ecological disintegration, novel infectious pathogens are likely to continue to (re)emerge (Arenas 372).

Our book strategically focuses on the intersection of pandemics with ecological breakdown. We choose this term rather than the more common "climate change" because "ecological breakdown" is more inclusive, encompassing other intersecting cataclysms such as the loss of "biosphere integrity" (Richardson et al.) often referred to as "the sixth mass extinction," as well as the wider disintegration of complex ecological systems (O'Connor et al.), and thus more accurately reflects the interdependent nature of the biosphere / "Terra" along with requisite responses to eco-breakdown. An exclusive or predominant focus on the "climate crisis," for instance, obscures the equally, if not more destabilizing, biodiversity crisis (Ceballos et al. 135966; WWF). This risks exacerbating the current situation, wherein the climate crisis has managed to gain such wide-scale public and government attention (Dunne), while biodiversity decline remains relatively unheeded outside of specialist and activist circles, and is still treated as separate from the climate crisis at the annual COPs.

It is also worth highlighting that great cataclysms and end-time events are not unique or exclusive to the present. Oppressive systems like colonialism, capitalism, patriarchy, and human supremacy have been ending worlds for many since at least 1492 (Ferdinand 26). Such vastly uneven spatial and temporal dimensions of apocalypse shed much-needed critical light on the "false globalisms" suggested by many (Eurocentric and Christian-centric) hegemonic Western discourses about "*the* apocalypse" (Lynch et al. 4). This is a pluriversal world of many worlds and ways of being, and likewise, many ends (Escobar 2). Thus, it is never the end of all possible worlds, but the end of "the world" as "we" have come to know it, of what we hold dear, followed by new possibilities for radical renewal and regeneration. Thus, end-time events also hold kernels of hope (McIntosh ix). Therein lies the transformative, emancipatory potential of utopian thinking and praxis, which thrives in the disjuncture between the hegemonic "market-liberal utopia" of our present (Žižek 38) and the innumerable possible worlds that might yet be brought forth through collective action.

Much classic and contemporary speculative and climate fiction, including H.G. Wells' *The Food of the Gods* (1904), Ursula Le Guin's *Always Coming Home* (1985), Margaret Atwood's *MaddAddam Trilogy* (2003, 2009, and 2013), the graphic novel-turned-film and series *Snowpiercer* (Lob et al. 2020), and Robinson's *Ministry for the Future* (2020), offers poignant observations on the potentially devastating, unforeseen ramifications of hubristic human bio-geoengineering feats. Other works more generally explore the potential existential as well as ethical ramifications of "human"-induced ecosystem global collapse. For instance, Elisabeth Sanxay Holding's short story "Shadow of Wings" presages wide-scale ecological collapse when birds mysteriously start disappearing from earth, generating a series of trophic cascades, including the gradual disappearance of insect life. The latter, in particular, threatens mass famine, which "means death, and plagues, riots, insanity and chaos…worse than earthquakes, volcanoes, hurricanes, tidal waves, because it's slower" (157). But, the potential end of other species matters not only because their absence would entail the end of us. The profound ethical weight of (particularly mass) species loss stems from the irrevocable severing of singular, intergenerational kinship relations (van Dooren, *Flight Ways* 27) forged over vast spatial-temporalities: "Death marks each time, each time in defiance of arithmetic, the absolute end of the one and only world…the end of the totality of what is or can be presented as the origin of the world for any unique living being, be it human or not" (Derrida 140). Much loss is yet to come as climate and biodiversity breakdown intensify, along with related cataclysms like novel zoonotic diseases.

Though there are no absolute ends, the loss of cherished kin is indeed irreversible. Novel forms and relations will come after, and one can resurrect hybrid bodies through de-extinction (Minteer). However, with the latter, what is brought back is not the person but a corpse devoid of the singular relations, memories, experiences, and desires that made them "them." Hence, Haraway's ethical injunction to honor our responsibilities to our terrestrial kin by "staying with the trouble" on terra (*Staying*), wherein we do our best to reckon with and stem the damage of colonial-capitalist ruination, (re)creating and nourishing spaces more conducive to multispecies flourishing (Chao et al.). Existing and novel viral and microbial symbionts will continue to be inextricable parts of this world-building, our companions as we navigate a challenging terrain ahead. In this dance with viruses, hurricanes, and innumerable other terrestrial agencies lies a vast ocean of possibilities. The latter are what utopian and dystopian studies seek to shed critical light on.

Keeping Hope Alive: Utopian and Dystopian Studies

The editors of this volume come from different pedagogical, cultural, and theoretical backgrounds. Rhiannon is a sociologist, Heather is an environmental social scientist and ethicist, Nora and Emrah are literary scholars, and Conrad is a literary scholar and author. Likewise, the chapters themselves are written by colleagues from different cultural, educational, and theoretical backgrounds, including literary studies; political science; urban studies; media studies; theater and performance studies; and community-based ethnography—although the volume in general leans toward the literary. While we hail from different disciplines, we all share an interest in the interdisciplinary field of Utopian Studies, and this interest acts as the theoretical substrate of the collection. The terminology of Utopia is 500 years old, invented as a neologism by Sir Thomas More as the title of his novel *Utopia* (1516), but most scholars agree the phenomenon it refers to is ancient, or even universally human, with notable examples such as Plato's *Republic* (ca. 375 BC) and *The Land of Cockaigne* (mid-fourteenth century). Utopias take many forms, especially literature, social theory, and lived experimental communities (Sargent), but we can also find utopian ideas and imagery in art, architecture, fairy tales, advertising, and medical textbooks—indeed, utopia pervades all spheres of life as the "principle of hope" (Bloch).

While utopias offer imaginative projections of better worlds and ways of being, dystopias extrapolate from the deficient present and offer projections of potentially nightmarish futures. Inherent within both are critique, imagination, and desire for the "better," or the queering and transgression of taken-for-granted borders and boundaries (Sargisson). Both utopias and dystopias share the function of mobilizing political affect—whereas utopias rouse hope and desire for a better future, dystopias trigger fear for the worst. Since the 1970s' wave of social movements and critique, many literary "critical utopias" tend to blend elements of both utopia and dystopia (Moylan).

The current approach to utopian and dystopian studies sheds light on the intricate relationship between these seemingly contrasting notions, which have traditionally been viewed as antonyms. The act of labeling an incident or a literary work as utopian or dystopian is inherently subjective, contingent upon the observer's perspective. This subjectivity not only complicates strict definitional boundaries but also gives rise to what Dunja Mohr aptly describes as the "intersection and comingling of dystopia and utopia" (64), which perhaps highlights how utopia—the so-called perfect world—is unattainable. In this context, one might even argue for treating utopia and dystopia as potential synonyms, ultimately eroding the traditional "distinctions between eutopian and dystopian spaces" (Marks 161) and underscoring a necessary linkage to societal realisms (Scott 43–44, 47–48). The complexity deepens under the pressures of contemporary challenges, notably exemplified by ecological breakdowns and other crises. The urgency for a non-dichotomous approach to understanding utopia and dystopia becomes even more pronounced. Such an approach is crucial for revealing the multidimensional nature of incidents, phenomena, and situations in the face of these crises, which is discussed throughout the chapters in this edited collection. For instance, the ongoing ecological crisis not only challenges our traditional notions of a utopian existence but also blurs the lines between utopian aspirations and dystopian realities, as anthropocentric practices have resulted in dystopian experiences. Similarly, the dystopian experiences of pandemic and other virulent infection, which can lead to illness, death, and the destruction of community, can also lead (e.g., through the transformation of social and physical infrastructures) to chances for change and living-otherwise. This is not to diminish the toll of pandemics, but rather to highlight that reality is not so black-and-white. Hope flourishes amid uncertainty, and plays a crucial function in both utopianism and dystopianism. As Patrick D. Murphy astutely notes, "Sometimes hope and potential do more to break inertia" (6). The focus on hope, which permeates this volume, imbues the study of utopia and dystopia with a dynamic dimension. This emphasis prompts a profound exploration into how these genres navigate and respond to the intricate complexities of the contemporary world. By delving into the potential transformative power of hope, our volume seeks to enrich the understanding of utopian and dystopian narratives, shedding light on their nuanced engagement with the challenges and uncertainties of the present era—particularly as they are, and have been, transformed by both global health crises and environmental emergency.

Thinking through the transformative power of hope brings us to consider the role of utopias in producing social change. Indeed, it is probably fair to say that, as editors, our foremost investment in this project is our shared dissatisfaction with the status quo. Beyond this, we share desires for something better, and to play a role in exploring and shaping what that might be. As scholars of utopia, we all believe in the potential of cultural imaginaries and the utopian impulse in providing critical exegesis and producing social change. While we cannot reach out and touch utopia, and while utopias are often portrayed as "impossible" or unreachable horizons, they play an essential role in social change: "they hold up a mirror (to the flaws of the present) and they inspire (saying 'things could be so much better')" (Sargisson 8). Thinking beyond the spatial, temporal, and conceptual boundaries of the decaying status quo is essential for building beyond the current "eco-dystopian" era of pandemics, extinctions, and ecological collapse. Utopia is both critical and visionary (Moylan and Baccolini), and the utopian model of social change involves disentangling positive visions from reactionary and reactive aspects of social/political action, while acknowledging that critique and social action are inseparable and never-ending processes.

Through the figure of pandemics, then, the focus in this book is on utopia-as-process. The foremost critique leveled against utopia since time immemorial, and most famously by Karl Popper, equates utopias with centralized leadership, impossible perfection, and totalitarianism. Popper argued that all attempts to impose a fixed utopian blueprint upon imperfect and diverse humankind are incompatible with Liberal freedoms and democracy, which requires reformist, "piecemeal" change (Popper 177). Thus, later (neo-)liberals such as Francis Fukuyama would proclaim that in liberal democracy we had reached "the end of history" (Fukuyama), while the British Prime Minister Margaret Thatcher famously intoned that "there is no alternative" (Pohl 3). The Liberal view assumes desires are individualized or confined to the private family, or to market relations, and politics requires a monopoly of force to protect this private realm. Collective and socialist utopias are seen to be doomed to failure and totalitarianism—the assumption being that they claim to be conflict-free (which is not the case) and to require a top-down and totalitarian imposition. This view conveniently ignores how violent, top-down, and authoritarian neoliberal Structural Adjustment Programs have been, which themselves were based on an implicit utopian vision of an individualized, friction-free free market utopia. As David Graeber argued,

> Stalinists and their ilk did not kill because they dreamed great dreams … but because they mistook their dreams for scientific certainties … In fact all forms of systemic violence are (amongst other things) assaults on the role of the imagination as a political principle. (10–11)

Authoritarianism is a problem of centralized and unequal power and resources, and not with utopian dreaming! Thus, many utopians favor forms of social change that are experimental and prefigurative, proceeding by immanently enacting organization and relationships, building common infrastructures, and coproducing knowledges at a grassroots level that are being sought in the "future" society (Boggs; Raekstad and Gradin). Prefigurative politics and its role in social change is a concern at the forefront of the utopian visions of social movements and social experiments, such as those covered in the chapters by McKnight, Kafara, and Grześkiewicz and Boschen, yet we also find portrayals of prefigurative movements in some of the literary utopias covered in chapters by Leggatt and Käkelä, as well as in the arts-based practices covered by Breemen. Whether they are articulations of imagination or of human organizational practice, prefigurative politics can serve a consciousness-raising function, building knowledge of previously invisibilized and silenced structural oppression starting from the embodied intersectional experiences of oppressed/marginalized people, while allowing them/us to explore new and diverse visions of what it might mean to build lives beyond this. Yet often prefigurative politics and its concerns can seem all too human. Our desire in this volume has been to move beyond the human in imagining alternative futures—to prefigure and explore imaginaries around non-exploitative relations with nonhuman nature.

We find that literatures on new materialism, posthumanisms, and multispecies conviviality are invaluable for our conceptualizations here. Posthumanisms variedly critique and deconstruct rigid dualisms between humans and other-than-human entities of various kinds. Critical posthumanism is especially interested in dismantling the ethico-ontological foundations of human supremacy, its hierarchical and dualistic constructs, while avowing the inescapably relational nature of terrestrial life and gesturing toward affirmative ethical modalities of multispecies conviviality (Braidotti; van Dooren and Rose). This drive to dismantle dualisms and dichotomies in favor of more complex and interconnected ways of understanding and being in the world is increasingly reflected in various types of cultural and social production: "As contemporary science reveals the ways our bodies exist as shifting and sharing ecosystems rather than isolated and inviolate objects," explain de Bruin-Molé and Polak, "contemporary culture unveils a seemingly endless series of new ways to be, to behave and to belong alongside other bodies" (3). All of the chapters in this collection variedly gesture toward the ultimate untenability of static and rigid borders—conceptual, national, between the purportedly sovereign "Self" and the "other," and between humans and other species. Borders, rather, are always contestable, liminal, and fluid, requiring continuous (re)negotiation. Ecological breakdown, like pandemics, profoundly unsettles notions of human supremacy and autonomy. The COVID-19 pandemic reminds us of the powerful agency of our viral and microbial symbionts, who make up most of our DNA, immune systems, guts, and skin (Gilbert et al.), and who play significant roles in influencing global biogeochemical cycles, such as oceanic carbon sequestration (Danovaro et al.). Herein, too, lie boundless utopian possibilities for reconfiguring better relations with others within and beyond the ruins of capitalism (Tsing; Tsing et al.).

Themes and Chapter Summaries

The book is divided into four thematic parts: Monsters and Monstrosity, Intersectional Critique, Eco-Justice and (More-Than-Human) Mutual Aid, and Creative Resistance and Utopian Glimmers. These sections are loose categories rather than binding boxes, and many of the chapters' themes and ideas interconnect and overrun their boundaries. There are entanglements within, between, and beyond these sections that we do not seek to limit by positioning them in one category or another. Nor are the chapters organized in any linear fashion; one can just as well start at the end, beginning, or somewhere in the middle. We have taken instances of overlap and interconnection as ways to bridge the transitions between sections, with an eye toward invoking the porosity and permeability of boundaries toward which the book as a whole gestures. We encourage readers to look toward what connects the sections to one another rather than to what separates them, to read for the tentacular or rhizomic interweavings. At the same time, utopia should be filled with difference and diversity, a cacophony of voices and ways of being and knowing united by a similar drive toward building better futures. We have encouraged multiple nodes and heterogeneous lines of flight, and, as one must do in such an interdisciplinary endeavor, trusted our authors where their expertise diverges and/or exceeds our own.

The book's organization is meant to form a kind of theoretical crescendo, a building up and opening outward. It begins with Monsters and Monstrosity and the consequent questioning of restrictive binaries, then opens outward to not only question binaries but also acknowledge (and respect) multiple differences in Intersectional Critique, then goes beyond acknowledgment to collaboration, collective action, and mutual aid in Eco-Justice and (More-Than-Human) Mutual Aid, and finally opens up fully in Creative Resistance and Utopian Glimmers to look toward what might be possible beyond aid through creativity and utopia-as-praxis. Each section finds resonances in the others while focusing on its own themes, which are elaborated upon below.

Monsters and Monstrosity

An oft-quoted definitional feature of utopia is that it functions as "the education of desire" (Abensour cited in Levitas 141; Davis 484–85). Monsters, similarly, serve a pedagogical function: rousing political affects such as fear and horror. Tyson E. Lewis and Richard Kahn describe Marx's "science of history" (17) as a bestiary (a compendium of monsters), with capitalism analogized through his works as various occult figures, including a blood-sucking vampire (19–22), a shape-shifting werewolf hungry for surplus labor (22–24), and as a veiled Medusa (24–27), whose contradictions cannot be held together in a totality. Despite the seeming rifts between historical materialism and new materialism, perhaps even our most scientific historical materialist cannot help but be posthuman after all, using monsters and posthuman entities as metaphors to illustrate his points pedagogically.

This part comprises three chapters: Tânia Cerqueira's "'In the woods the Tox is still wild': The EcoGothic in Rory Power's *Wilder Girls*," Ujjwal Khobra and Rashmi Gaur's "The Human/Un(human): Monster, Ecophobia, and the Posthuman Horror(scape) in Dibakar Banerjee's 'Monster,' *Ghost Stories*," and Timothy S. Murphy's "A Scourge Even Worse Than Disease: Richard Matheson's *I Am Legend* as Pandemic Political Allegory." Each of these chapters addresses texts in which a contagion causes those infected to become monsters, whether those be posthuman hybrids with gills and talons, hairy four-legged creatures with a taste for human flesh, or living and undead vampires. The invocation and literalization of monstrosity in these various texts works to question human/nonhuman boundaries, as well as to put under pressure various ways of relating to and with other humans.

In her chapter, "'In the woods the Tox is still wild': The EcoGothic in Rory Power's *Wilder Girls*," Tânia Cerqueira explores Rory Power's young adult novel, *Wilder Girls* (2009), through the lens of the ecoGothic. The ecoGothic, Cerqueria explains, is a mode that utilizes Gothic tropes—particularly that of the uncanny, or *unheimlich*—within an environmentally minded context to highlight issues of ecological devastation and anthropogenic climate change. Drawing on theories of posthumanism and ecophobia—theories also taken up in the subsequent chapter by Khobra and Gaur—Cerquiera argues that Power's tale of mutated, posthuman bodies caused by the highly contagious "Tox" demands a perspective shift on the part of its readers from ecophobia to ecocentrism. The protagonists, all young women, "find power within their entanglement with an ecoGothic nature (38)," and the narrative thereby "urges a nonhuman turn (38)" that recognizes that the "monstrosity" of human entanglements with nature is not something to reject outright out of fear. This chapter sets the stage for subsequent chapters' exploration of contamination, including Eleonora Rossi's work on porosity, ecofeminism, and the Blue-Green humanities in the next section.

In their chapter, "The Human/Un(human): Monster, Ecophobia, and the Posthuman Horror(scape) in Dibakar Banerjee's 'Monster,' *Ghost Stories*," Ujjwal Khobra and Rashmi Gaur discuss the "Un(human)/Monster as a posthuman reconfiguration through an extensive analysis of Dibakar Banerjee's short film, "Monster," part of the horror anthology, *Ghost Stories* (2020). In their critical discussion, Khobra and Gaur draw on Simon Estok's ecophobia and Rosi Braidotti's politics of otherness to reveal the need to adopt an "inter-relational" approach. Their analysis focuses especially on cannibalism in the film, which becomes part of a new social order, and which, historically, has worked to destabilize understandings of the "human," and therefore the dichotomy between the "human" and the "nonhuman other." Through their discussion of the film, Khobra and Gaur highlight the necessity of deconstructing the "human" and questioning our relationship with nature and the non/inhuman world, as well as the need to challenge conventional anthropocentric limitations in order to have another possible future, which involves a different relationship with the planet.

In the subsequent chapter, "A Scourge Even Worse Than Disease: Richard Matheson's *I Am Legend* as Pandemic Political Allegory," which serves as a bridge from this section to the next on Intersectional Critique, Timothy S. Murphy reads Richard Matheson's 1954 vampire novel, *I Am Legend*, as political allegory, invoking both reactionary, anti-democratic movements and racialized questions of civil rights, particularly in light of the COVID-19 pandemic and its political repercussions and reverberations. After a bacterial pathogen causes widespread infections, which turn humans either into living vampires or undead ones, the lone human protagonist, Neville, "clings to his inevitably waning power more furiously, violently, and blindly as it slips away" (156), highlighting the question of what happens when the majority-become-minority fails to adapt to new (political) circumstances. The vampires of this novel are not mindless blood-suckers beholden to their urges, but rather savvy political actors more prepared than humans for the coming of a new (posthuman) world. Neville's reactionary opposition, Murphy argues, through reference to a variety of real-world politics and political movements, highlights the paranoia, anxiety, and violence often triggered by the establishment of new types of citizen-subjects, whose very existence threatens traditional, entrenched power structures.

Intersectional Critique

The prominent critical race theorist Kimberlé Crenshaw coined the term "intersectionality" in an effort to capture the myriad overlapping social forces and social identities through which power and disadvantage legitimize and express themselves. As all lenses and positionalities are necessarily partial, they generate blind spots that can easily remain overlooked without continuous attention and critical reflection. The nuanced lens of intersectionality, crucially, helps widen our fields of vision—our analytical, diagnostic and prognostic framings—so as to take account of a broader range of factors, structural inequalities, and modes of exploitation, and of how they overlap and often compound one another to perpetuate and create new modes of exploitation and vulnerability. Malcolm Ferdinand, in his poignant recent work *Decolonial Ecology: Thinking from the Caribbean World* (2021), sheds important light on the "modern double fracture" at the heart of Western environmental movements that commit the fatal error of depicting ecological, animal, BIPOC, and feminist causes as "fundamentally separate issues" (215). Indeed, environmental movements and organizations in the Global North have long been marked by a distinct lack of racial, class, and other forms of diversity both in composition and narrative framings (Bell and Bevan; Thomas), including Transition Towns (Grossmann and Creamer), Greenpeace (Harter) and more recently Extinction Rebellion (James and Mack).

Some writers in black feminist traditions (Nash) emphasize the importance of affective political modalities of love (Hardt and Negri) that embrace difference and radical possibility. Some have highlighted the potential for intersectionality's identitarian politics to be taken up merely as a tool for "difference management" within a wider mantra of liberal multiculturalism (Puar, "Queer Times" 128). More problematically, as Jasbir Puar observes, this mantra often "colludes with the disciplinary apparatus of the state—census, demography, racial profiling, surveillance—in that 'difference' is encased within a structural container that simply wishes the messiness of identity into a formulaic grid" (128). In place of intersectionality, Puar advocates theorizing "assemblage," which "underscores feeling, tactility, ontology, affect, information" and heterogeneity (*Terrorist Assemblages* 215). Black feminist affective politics fundamentally advocates building communities united in diversity rather than homogeneity (Nash 451). Similarly, a central pillar of Martin Luther King Jr.'s politics of love and hope is an "inescapable network of mutuality" (King Jr. 73) that binds *all* earthly beings together. These vast networks of mutuality also hold immense potential for radical transformations of our unjust and unsustainable status quo, if mobilized through collective action in service of liberated futures. However, for King, the "cancerous egotism" of Western (capitalist) societies (King Jr. 72) has often worked to obscure the myriad interdependencies that sustain us.

Intersectional critiques and framings firmly grounded in "forms of collective sentiment" or affect, mobilized by love-politics that embraces difference rather than mere identity (Nash 451), thus harbor considerable utopian potential for radically upending the structural drivers of contemporary socio-ecological disintegration. Single frames that focus only on "environmental," "human rights," "race," or "gender" issues, as though they could ever be so neatly disentangled from one another, override the irreducible complexity and interdependent nature of the world and justice struggles. To quote intersectional activist and scholar Audre Lorde, there is "no such thing as a single-issue struggle because we do not live single-issue lives" (42). Relatedly, Ferdinand calls for interspecies alliances that unite the concerns of feminist, environmental, labor, and BIPOC activists into a common problem with similar origins (215), if we are to effectively resist forces of colonial-capitalist ruination and (re)create conditions for a mutual flourishing within and after the Capitalocene (Moore, *Capitalism in the Web of Life*).

In a related sense, Matthew Leggatt's chapter, "The Reconfiguration of Post-Pandemic Society in *Severance* and *The Last of Us*," sheds light on the heterogenous networks of mutuality often undermined and exploited by global capital, yet which also always hold unexhausted potential for novel post-capitalist (re)configurations. Leggatt elucidates the rhizome as both a metaphor for horizontal flows of capital and apparently (but not actually) disorganized capital (determined by deregulation) that masks embedded systems of inequality, and creates the conditions to be harnessed by "malevolent forces"—a dynamic exacerbated by the shock and seeming randomness of COVID-19. At the same time, "fungal imaginaries" exhibit a utopian aspect, connecting humans in ways "that may eventually be transformational" (70) as horizontalist social movements. Matthew coins the ingenious and amusing "fungal capital" to describe a corrupt economy that appears arbitrary because its connections between nodes are not immediately visible, existing below the surface. They are to be uncovered by "following the money," as with the PPE (personal protective equipment) scandal where the UK government gave out lucrative contracts to some of its wealthiest friends to manufacture masks and other protective equipment (71). The utopian aspect of this dynamic, and its intersection with climate change, embodied in "new fungal relationships and communities" (71), is explored through an analysis of Ling Ma's *Severance* and HBO's adaptation of Naughty Dog's *The Last of Us*.

Benjamin Burt's chapter, "Five Hundred Years of Plague: Indigenous Apocalypse in Joca Reiners Terron's *Death and the Meteor*," examines the representation of apocalypse in Brazilian novelist Joca Reiners Terron's 2019 text *Death and the Meteor* (*A morte e o meteoro*). The chapter contextualizes the book in the history of colonialism, arguing that, for many Indigenous Amerindians, the apocalypse has already happened, as the plagues brought by the colonizers "attained pandemic, world-annihilating proportions" (81). Burt continues to explore the themes of cataclysm and post-apocalyptic renewal through Terron's novel, wherein the surviving members of the Kaajapukugi tribe, already on the edge of extermination after decades of epidemics and environmental degradation, consume the last of their sacred hallucinogen and commit ritual suicide. This apparently ushers in the end of the world via a meteor strike, yet there is an ambiguous element of post-apocalyptic hope via the tribe's belief in eternal return.

In her chapter, "Corruption and Cleansing: An Eco-Feminist Approach to the Nature/Culture Dichotomy in Naomi Novik's *Uprooted*," Sara González Bernárdez applies a posthumanist approach to Naomi Novik's 2015 novel *Uprooted* through an extensive discussion of the nature-culture dichotomy illustrated in the narrative. By drawing on influential theoretical sources and arguments by Rosi Braidotti and Val Plumwood such as relationality, interdependence, and ethics of affirmation, and by analyzing significant characters such as Agnieszka, Sarkan, and the Wood-queen in *Uprooted*, González Bernárdez invites us to re-evaluate our perceptions of the nonhuman world in order to reassess and transform our ingrained perceptions so that we may become more and more aware of our interdependence, foster respect for the nonhuman world, and shape the future accordingly.

The last in this part, Eleonora Rossi's chapter, "Through Currents of Contamination: The Failure of Immunizing Insularity in Sophie Mackintosh's *The Water Cure*," features a feminist analysis of the porosity of bodies, particularly classic interpretations of the porous female body as monstrously "other," as explored in Sophie Mackintosh's *The Water Cure* (2018). In the novel, porosity and fluidity are subordinated to superior "male" attributes of (supposed) impenetrability and rigid binaries. The novel's revelation that the island housing its protagonists is actually a peninsula, argues Rossi, underscores the ultimate interconnectedness and permeability of all terrestrial things, rendering complete isolation impossible. In this entangled, terrestrial existence, there are no islands (Donne 87).

Although not always explicitly drawn upon, intersectional critiques and alliances surface throughout many of the chapters in this collection, such as Hanna Grześkiewicz and Marleen Boschen's chapter, "(Un)Caring Borders: More-Than-Human Solidarities in the Białowieża Forest," in the next part.

More-Than-Human Mutual Aid and Eco-Justice

During the COVID-19 pandemic, the term "mutual aid," which had previously been the preserve of anarchist and other radical leftist movements, entered mainstream usage, with even conservative politicians encouraging its use as a concomitant to "Big Society" and a means to encourage unpaid social labor to mop up the failures of the austere neoliberal state (Kruger; Mould et al.). Some academic outputs on mutual aid that arose from the pandemic mirrored the government and media discourse, attempting to re-historicize and de-politicize mutual aid, reframing it within the Liberal traditions of social capital, resilience, vulnerability and associationalism (Fraser et al.). Others paid homage to the roots of mutual aid in radical, leftist, working class, racialized, and LGBTQ+ mutualist movements and intellectual traditions, and emphasized the critical and radically transformative potential of such movements (Preston and Firth; Firth; Mould et al.). Nevertheless, the literatures on mutual aid and pandemics to date have been in the titular words of the philosopher Nietzsche, "Human, All Too Human." The relevance of Nietzsche extends beyond catchy titles, as one of the central concerns of his body of work was the way in which humans relate to their animal beings, and indeed Peter Kropotkin himself, the anarchist philosopher who originally popularized the term "mutual aid," did not forget about the animals in his conceptualization, which was formulated as a response to the Darwinists and Social Darwinists of his time. In fact, the first two chapters of *Mutual Aid* draw on studies of animals—especially ants, bees, birds, rodents, wolves, monkeys and lions (Kropotkin 12–51). Kropotkin draws on these studies of animal behavior, alongside studies of peasant and Indigenous groups (which he refers to in somewhat dated language) to show that other-than-human animals and people often cooperate, not only within but also between species, even in conditions of scarcity and emergencies. Co-operation can be just as important as self-protection, and the struggle for survival is often against circumstances rather than between individuals (Kropotkin 14). The anarchist theorist Max Stirner is also worth a mention here, as through his idea of the "Un-Man," he connects a critique of humanist discourse, which prefigures posthumanism, to a valorization of the monstrous—those parts of ourselves that do not fit within the idealized political/capitalist concept of "man" and yet that are essential for relating authentically in a "union of egoists," akin to a mutual aid relationship between equals who refuse to be dominated by alien ideologies (179). Erica Cudworth and Stephen Hobden draw on the work of Norbert Elias, a theorist of civilizing process, to show how the oppression of our animal selves, and repression of our own instincts and desires, leads to a particular grouping of humans (those elite who see themselves as more "civilized") being enabled to separate themselves from other humans, other animals, and nature, and therefore to exert dominance, using those beings deemed "uncivilized" as mere objects to mobilize in meeting the "needs"/desires of the civilized. Cudworth and Hobden argue that "those groupings who make the claims to be most separated from nature are those posing the gravest ecological threats" (746). Going back to Kropotkin's original formulation of mutual aid, then, we want to highlight ways in which chapters in this volume think through the means through which species can unite and work together in mutual aid against the domination and destruction being wrought by a small elite group of humans.

The idea of More-Than-Human Mutual Aid in the contemporary moment is, we believe, a novel concept that was inspired by some of our chapter authors who seek to reconfigure relations of mutuality, justice, and care beyond species boundaries and indeed at times extending to the entire ecosystem, thinking through how our relations within the web of life are symbiotic and mutually supporting, and how political action around eco-justice is not done *for* the environment, or other species, but rather with and within our mutually sustaining relations. Eco-justice is a utopian movement toward a world wherein none—neither humans, other animals, nor rivers and plants—are rendered disposable or sacrificial. Haraway's concept of the Chthulucene similarly evokes such a world, wherein "intense commitment and collaborative work and play with other terrans" across multiple spatial-temporalities ("Anthropocene" 160) can help us recreate spaces of refuge, promote multispecies flourishing, and more generally "live and die well" in the terrain ahead.

In the first chapter of this part, "Dystopian Prohibitions and Utopian Possibilities in Edmonton, Canada, at the Onset of the COVID-19 Pandemic," Rylan Kafara discusses the links between Indigeneity and eco-justice in the context of a relief encampment movement that arose in several cities in Canada in the early months of the COVID-19 crisis. He paints a picture of capitalism and gentrification as extractive and dispossessing, anti-human forces. Yet there is a utopian side to his argument, of course: he shows how a social movement of Indigenous peoples created spaces for "harm reduction, mutual aid, community care" based on the utopian ideal of "coming home" enacted through the "relief encampments," including one named Pekiwewin (Nêhiyawak/Cree for "it is sheltered") (127). He explores the theme of temporality, which also runs through many other chapters, showing how prefigurative spaces in which inhabitants live the future they would like to see in the present illustrate utopian alternatives to racial capitalist settler coloniality and proffer hope for a sustainable future.

Claire P. Curtis' chapter, "Affiliation as Environmental Justice in Three Climate Novels," the next in this section, focuses on environmental justice in three climate fiction novels, examining how each of the novels generates different understandings of a just community in their post-pandemic milieus. In her readings of Leigh Richards' *Califia's Daughters* (2004), Emily St. John Mandel's *Station Eleven* (2014), and James Bradley's *Clade* (2017), Curtis relies on Martha Nussbaum's concept of affiliation—that is, "living with and toward others" (Nussbaum 77). Curtis argues that pandemics pose "a special and direct challenge to affiliation," as they prompt "a generalized anxiety about whether and how we will be able to live together" (135). In reading the three texts, she looks toward how they depict their respective characters' capabilities for affiliation, and how that, in turn, functions within, is informed by, and informs the ethos of their communities, particularly against the backdrop of their respective histories of world-changing pandemics. Ultimately, she argues through her comparative analysis, there is a need to look beyond enclosed, isolated communities, and to stop imagining that solutions will simply arise from crisis. Instead, Curtis argues, climate fiction can show us the need to preemptively and proactively cultivate environmentally just frameworks that reach out toward the greater (human and nonhuman) world, instead of folding inward into protective—but ultimately unsustainable—isolationism.

The subsequent chapter, "'A vortex of summons and repulsion': The Productive Abject, Posthumanisms, and the Weird in Charles Burns' *Black Hole* (2005)," authored by Benjamin Horn, Jayde Martin, and Alice Seville, examines the titular graphic novel and its timely themes of boundary porosity and dynamism, particularly in a post-COVID-19 world wherein the powerful agency of minute other-than-human entities has become starkly apparent. As this chapter argues, the characters in *Black Hole*, plagued by and striving to grapple with a sexually transmitted virus that produces strange other-than-human effects, come to learn the power of viral entities in undermining such illusory conceptual dualisms as human/nonhuman, inside/outside, life/non-life that have come to dominate Western cultures. Rather than depicting the virus as an external invader or intruder threatening to undermine the "human" characters' autonomy by altering their biology and physiology, as Seville, Horn, and Martin observe, Burns' *Black Hole* depicts the virus as inextricably interwoven in the teens lives. The boundaries of the human are revealed to be "permeable at all levels" (154). Thus, the authors conjecture that the initially revolting human-viral encounter, which subverts self/other distinctions, ultimately harbors a potentially transformative "productive abjection" (158), opening the space for a counter-hegemonic politics of multispecies conviviality.

Closing out the section, Hanna Grześkiewicz and Marleen Boschen's chapter, "(Un)Caring Borders: More-Than-Human Solidarities in the Białowieża Forest," considers the entanglements of COVID-19, the humanitarian crisis of people on the move, and the stag beetle in an examination of the ways in which policies are enacted upon life in Europe's last remaining primary forest, the Białowieża Forest. The forest is arbitrarily crossed by the Polish-Belarusian border, creating utopian and dystopian narratives of the forest and the biodiversity and life it sustains, and the destructions wielded by the policies of capitalistic nation states that aim to "guard" it through states of emergency and necropolitics. They also consider utopian aspects to this dynamic: not only the forest's primal biodiversity but also the resistance and protests against the logging and the bordering regimes of the humanitarian crisis. The authors theorize a "denial of care" by authorities, which is resisted by anarchist activists "providing people on the move with water, fresh clothes, SIM cards, powerbanks, food and warm drinks, as well as providing first aid" (169).

Creative Resistance and Utopian Glimmers

Our collection is a cultural product, and its chapters analyze cultural representations of pandemics and ecological breakdown, understood broadly to encompass a wide variety of expressions from fiction to social movements to theater. The chapter authors are individuals and small collectives, usually writing from an academic perspective. Our collated writings are a form of political and politicized praxis—as we have stated previously, the key function of utopian thought is to transform consciousness and society. Nevertheless, there is also an important and long tradition of anarchist and autonomist collective praxis that operates in the realm of culture beyond representation, which places emphasis on collective prefigurative action and Do-It-Yourself (DIY) culture and politics (McKay). This sometimes involves reclaiming ideological or territorial space from state and capital to experiment and build new political cultures drawing on a non-hierarchical organizing and DIY political and ecological ethos, with an emphasis on creating something new, autonomous, and beyond capitalist extractivism. Some of our chapters provide not only a utopian analysis of cultural products and projects but also signal toward these other kinds of posthuman collective practice, creative resistance, and utopian transformations in the realm of culture and DIY politics.

The chapters in this section variedly gesture toward the wealth of utopian possibilities that persist, indeed proliferate, within and against the omnicidal (Ghosh) wreckage of (neo)colonial-capitalism. The late multispecies ethnographer Deborah Rose Bird refers to the concept of "shimmer" in relation to the ancestral power of life (Rose Bird 54) that pervades and moves the biosphere, which she learned from her encounters with Aboriginal people of the Victoria River region of Australia's Northern Territory. Even amid the onslaught of late capitalist extractivism and its great unraveling of relationships and mutualities, when "all we love is being trashed," encounters with the enduring shimmer of our vibrant world hold radical utopian potential for (re)inspiring more attentive and caring relations with others (52). Indeed, as Rose Bird points out, the increasing dearth of life's shimmer ought to be understood not (only) as lack but as potential (55). As the growing absence of many other species in familiar places previously characterized by their notable presence sparks a renewed awareness of them, the absence of shimmering can incite a "fresh grasp" of the "awful disaster of extinction cascades" as the annihilation of life's potentiality (Rose Bird 55). Such encounters can, in turn, fuel powerful forms of creative resistance against oppressive and exploitative systems bent on the annihilation of diverse life forms. Our final four chapters explore human and more-than-human creations that arise out of the wreckage of entangled pandemics and ecological breakdown to signal to a world beyond mere critique of the structures that caused them. This includes arts-based practice, cultural resistance, and autonomous politics that make space for the agency of nonhuman animals and nature in co-creating the new.

Jari Käkelä's chapter, "'Preservation is an action, not a state': DIY Utopian Enclaves and Ways out of Post-Pandemic Surveillance Capitalism in Sarah Pinsker's *A Song for a New Day*," which begins Part 4, analyzes Sarah Pinsker's prescient 2019 novel, *A Song for a New Day*. The novel imagines DIY utopian enclaves that spring up in opposition to a dystopian, post-pandemic future of surveillance capitalism, anti-congregation laws, and highly controlled artistic and cultural production. These enclaves are organized around live music performance as a method of resistance, allowing the formation of communities and grassroots activist collectives. Käkelä explores themes of enclosure and openness in the novel, focusing particularly on the confluence of music and technology that is being negotiated in the novel's platform capitalist milieu. (It is important to note that the anti-congregation laws in the novel are no longer medically necessary, although the narrative that they are persists as a means of political control.) He argues that, with her novel, Pinsker is "singing her way out of dystopia" (187), showing how the emotional power of music can act as a catalyst to spur (successful) political, collective action against the seemingly immutable, oppressive social, political, and legal structures that strive to keep people alienated and apart in the service of capitalist consumption.

In the second chapter of this part, "Pandemic Dramaturgy: Co-Designing the Performance *Dying Together/Futures* with the COVID-19 Virus," Alice Breemen discusses the use of theater during COVID-19 using streaming, public spaces, and socially distanced theaters in between lockdowns. She recounts a series of plays with a focus on the iteration titled *Dying Together / Futures*, by theater company Building Conversation, which explores various events, including terrorist attacks; climate change; and COVID-19, in which "people, more-than-human entities, systems, and ideas died simultaneously" (190). Due to the physicality of the theater, the virus itself becomes a more-than-human actor in the play, both an "imaginary invited guest to play in the performance," as well as "an actual dangerous actor in the invisible 1.5-meter space between the participants" (195). Breeman understands this event using the concept of "onto-epistemological theatricality" and the Braidottian idea of "middle ground": "a space where participants can practice and play with relational entanglements that characterize the Anthropocene" (191). The methodology highlights relational entanglements in theater and offers tools to navigate the Anthropocene.

In the following chapter, Hülya Yağcıoğlu discusses literary depictions of plagues and pandemics, drawing on various theoretical sources such as Timothy Morton's "hyperobjects," Jane Bennett's "vital materiality," Bruno Latour's actor-network theory, and object-oriented ontology. Through an extensive reading of three significant works, namely Daniel Defoe's *A Journal of the Plague Year* (1722), Albert Camus's *The Plague (*1947), and Orhan Pamuk's *Nights of Plague* (2022), Yağcioğlu seeks to find the traces of the utopian impulse within these plague narratives. By drawing particular attention to the entangled relationship between the human and nonhuman world within the Anthropocene, this chapter highlights the potential constructive aspect of plagues and pandemics and the need to adopt the notions of acceptance, tolerance, and respect. The possibility of yet another utopian order in the aftermath of an unanticipated catastrophe finds its expression here through an analysis of these three seemingly different, but also highly similar, plague narratives.

In the last chapter of both the part and the volume, "World-Building Enactments of the School Strike Movements during the Pandemic: Reading Youth Climate Crisis Movements through a Micro- and Nano-Utopian Lens," Heather McKnight discusses the agency of young people fighting for their future through school strikes. Her chapter focuses on the School Strike Movement that took place during the COVID-19 pandemic, arising from the broader youth climate movement. She focuses on what she calls "micro-utopian" aspects of their protests: those aspects of direct actions oriented not only on resisting the old order but also on creating a new world in its shell through "small prefigurative enactments of a different world" (215). Thus, McKnight's chapter approaches the entanglements of climate change and pandemics from the perspective of the anxiety and pessimism of young activists in the face of dystopian climate collapse. Yet she also articulates the hopeful aspects of this intersection in resistance: "Resistance anticipates a solution even as it expresses despair" (215). Although the pandemic prevented the youth climate movement engaging in its established forms of protest (assemblies), the new context also enabled the emergence of new forms of protest through online modalities such as live streams, social media activity, and a digital strike. The transformation of spatialities is further articulated through the idea of a "nano-utopian" movement (221) that emerges through digital and local entanglement with other movements such as Black Lives Matter in responding to global incidents. McKnight argues the "nano-utopian" arises from within the entanglements of planned and unplanned actions, whose intersecting affects create new emergent prefigurative visions.

Conclusion: From Futurities to Futurisms

With the concept of "entangled futurities," we echo and expand upon futurisms scholarship such as that explored through Taryne Jade Taylor, Isaiah Lavender III, Grace L. Dillon, and Bodhisattva Chattopadhyay's concept of "CoFuturisms," which examines significant human cultural movements "both within science fiction and fantasy and also reach[ing] past the boundaries of speculative fiction" (1) to eschew the "hegemonic" (5) and "Global North" (1) positioning of the term "alternative," since "CoFuturist creations often do indeed offer alternatives, [but] are more than alternatives" (5). That is, though "futurisms" necessarily suggest the imperative and opportunity of forward, generative, and healing cultural movements, the chapters herein explore how, unless we strive for those positivisms and changes, our futurities may very well be overcast by increasingly problematic circumstances affecting not only humanity but also the health of the whole planet.

Our volume encapsulates and builds upon the internationality and interdisciplinarity of our 2021 conference. In doing so, it mirrors the "hope" of Taylor, Lavender III, Dillon, and Chattopadhyay in "allow[ing] future scholars and creatives to continue pushing past … boundaries and categorizations as we work together to build a better future" (6). *Utopian and Dystopian Explorations of Pandemics and Ecological Breakdown: Entangled Futurities* sketches the contours of potential better futures and reinforces their urgency by asking human perspectives to not only consider but also actively interact with nonhuman entities in concrete, complex, and intra-connected ways—highlighting how these relationships are always already a fact of life. There is never a time or place where we (humans) are without or apart from our terrestrial counterparts. We are born, live, and die together. Will the futures of human societies merely boil down to a series of dystopian "events" like the outbreaks and spreading of pandemics, or can we humans shift our utopian desire to interact in a meaningful way within the web of life—including pathogens and other potentially dangerous microbes—with which we are already and always entangled? We hope, with this framework and the chapters in this volume, to humbly encourage the latter. We hope to foster and continue the important work of moving from uncertain futurities to futurisms that enfold all forms of life on this living, breathing blue speck wobbling about the cosmos.

Works Cited

Abensour, Miguel-Hervé. *Les formes de l'utopie socialiste-communiste: Essai sur le communisme critique et l'utopie*. 1973. Université Paris 1 Panthéon-Sorbonne, PhD dissertation.

Arenas, Alberto. "Pandemics, Capitalism, and an Ecosocialist Pedagogy." *The Journal of Environmental Education*, vol. 52, no. 6, 2021, pp. 371–83.

Barad, Karen. *Meeting the Universe Halfway: Quantum Physics and the Entanglement of Matter and Meaning*. Durham, NC: Duke University Press, 2007.

Bell, Karen, and Gnisha Bevan. "Beyond Inclusion? Perceptions of the Extent to which Extinction Rebellion Speaks to, and for, Black, Asian and Minority Ethnic (BAME) and Working-Class Communities." *Local Environment*, vol. 26, no. 10, 2021, pp. 1205–20.

Bloch, Ernst. *The Principle of Hope*. Vols. 1–3. 1954. Cambridge, MA: MIT Press, 1986.

Boggs, Carl. "Marxism, Prefigurative Communism, and the Problem of Workers' Control." *Radical America*, vol. 11, no. 6, 1977, pp. 99–122.

Braidotti, Rosi. "Posthuman Critical Theory." *Critical Posthumanism and Planetary Futures*, edited by Debashish Banerji and Makarand R. Paranjape. New Delhi: Springer, 2016, pp. 13–32.

Carrington, Damian. "Dramatic Climate Action Needed to Curtail 'Crazy' Extreme Weather." *The Guardian*, 28 August 2023, https://www.theguardian.com/environment/2023/aug/28/dramatic-climate-action-needed-curtail-extreme-weather.

Castle, Nora, Heather Alberro, Emrah Atasoy, Rhiannon Firth, and Conrad Scott. "Symposium: Living in the End Times." *SFRA Review*, vol. 51, no. 2, 2021, https://sfrareview.org/vol-51-no-2-spring-2021/.

Ceballos, Gerardo, Paul R. Ehrlich, and Peter H. Raven. "Vertebrates on the Brink as Indicators of Biological Annihilation and the Sixth Mass Extinction." *Proceedings of the National Academy of Sciences*, vol. 117, no. 24, 2020, pp. 13596–602.

Chao, Sophie, Karin Bolender, and Eben Kirksey, editors. *The Promise of Multispecies Justice*. Durham, NC: Duke University Press, 2022.

Crenshaw, Kimberlé Williams. "Mapping the Margins: Intersectionality, Identity Politics, and Violence against Women of Color." In *The Public Nature of Private Violence*, Fineman, Martha Albertson. New York: Routledge, 2013, pp. 93–118.

Cronon, William. "The Trouble with Wilderness; or, Getting Back to the Wrong Nature." *Uncommon Ground: Rethinking the Human Place in Nature*, edited by William Cronon. New York: W. W. Norton, 1995, pp. 69–90.

Cudworth, Erika, and Stephen Hobden. "Civilisation and the Domination of the Animal." *Millennium*, vol. 42, no. 3, 2014, pp. 746–66.

Danovaro, Roberto, et al. "Marine Viruses and Global Climate Change." *FEMS Microbiology Reviews*, vol. 35, no. 6, 2011, pp. 993–1034.

Davis, Laurence. "The Past and Future of Utopian Studies." *Utopian Studies*, vol. 34, no. 3, 2024, pp. 478–88.

de Bruin-Molé, Megen, and Sara Polak, "Embodying the Fantasies and Realities of Contagion." *Embodying Contagion: The Viropolitics of Horror and Desire in Contemporary Discourse*, edited by Sandra Becker, Megen de Bruin-Molé and Sara Polak. Cardiff: University of Wales Press, 2021, pp. 1–12.

Dillon, Grace L., editor. *Walking the Clouds: An Anthology of Indigenous Science Fiction*. Tucson: University of Arizona Press, 2012.

Donne, John. "Meditation XVII." *Devotions upon Emergent Occasions*. 1624. Edited by Anthony Raspa. Montreal: McGill-Queen's University Press, 1975, p. 87.

Dunne, Daisy. "Video: Why Does Nature Loss Receive Less International Attention Than Climate Change?" *Carbon Brief*, 21 June 2022, https://www.carbonbrief.org/video-why-does-nature-loss-receive-less-international-attention-than-climate-change/.

Escobar, Arturo. *Pluriversal Politics: The Real and the Possible*. Durham, NC: Duke University Press, 2020.

Farrier, David. *Anthropocene Poetics: Deep Time, Sacrifice Zones, and Extinction*. Minneapolis and London: University of Minnesota Press, 2019.

Ferdinand, Malcom. *Decolonial Ecology: Thinking from the Caribbean World*. Cambridge: Polity Press, 2021.

Firth, R. *Disaster Anarchy: Mutual Aid and Radical Action*. London: Pluto Press, 2022.

Fraser, Timothy, Courtney Tan, and Daniel P. Aldrich. "Won't You Be My Neighbor? Uncovering Ties between Social Capital and COVID-19 Outcomes at Local Levels." 18 February 2021. SSRN, http://dx.doi.org/10.2139/ssrn.3788540.

Fukuyama, Francis. *The End of History and the Last Man*. 1992. New York: Simon and Schuster, 2006.

Ghosh, Amitav. *The Nutmeg's Curse: Parables for a Planet in Crisis*. London: John Murray, 2021.

Gilbert, Scott F., Jan Sapp, and Alfred I. Tauber. "A Symbiotic View of Life: We Have Never Been Individuals." *The Quarterly Review of Biology*, vol. 87, no. 4, 2012, pp. 325–41.

Graeber, David. *Fragments of an Anarchist Anthropology*. Chicago: Prickly Paradigm Press, 2004.

Gross, Lawrence W. "Cultural Sovereignty and Native American Hermeneutics in the Interpretation of the Sacred Stories of the Anishinaabe." *Wicazo Sa Review*, vol. 18, no. 2, 2003, pp. 127–34.

Grossmann, Mena, and Emily Creamer. "Assessing Diversity and Inclusivity within the Transition Movement: An Urban Case Study." *Environmental Politics*, vol. 26, no. 1, 2017, pp. 161–82.

Haraway, Donna J. "Anthropocene, Capitalocene, Plantationocene, Chthulucene: Making Kin." *Environmental Humanities*, no. 6 (1), 2015, pp. 159–65.

———. *Staying with the Trouble: Making Kin in the Chthulucene*. Durham, NC: Duke University Press, 2016.

Holding, Elisabeth Sanxay. "Shadow of Wings." 1954. *Nature's Warnings: Classic Stories of Eco-Science Fiction*, edited by Mike Ashley. London: British Library Publishing, 2020, pp. 145–74.

Hovanec, Caroline. *The 1918 Influenza Pandemic in Literature and Memory*. 2009. Vanderbilt University, Master's Thesis.

Hulme, Mike. "Four Meanings of Climate Change." *Future Ethics: Climate Change and Apocalyptic Imagination*, edited by Stefan Skrimshire. London: Continuum, 2010, pp. 37–58.

IPCC. "Headline Statements." *AR6 Synthesis Report*, 2023, https://www.ipcc.ch/report/ar6/syr/resources/spm-headline-statements/

James, Diana, and Trevor Mack. "Toward an Ethics of Decolonizing Allyship in Climate Organizing: Reflections on Extinction Rebellion Vancouver." *From Student Strikes to the Extinction Rebellion*. Cheltenham: Edward Elgar Publishing, 2020, pp. 32–53.

King Jr., Martin Luther. *A Gift of Love*. London: Penguin Classics, 2017.

Klein, Naomi. "Let Them Drown: The Violence of Othering in a Warming World." *London Review of Books*, vol. 38, no. 11, 2016, p. 12.

Kropotkin, Peter. *Mutual Aid: A Factor of Evolution*. 1902. London: Freedom Press, 1987.

Kruger, Danny. *Covenant: The New Politics of Home, Neighbourhood and Nation*. London: Swift Press, 2023.

Lawler, Odette K., et al. "The COVID-19 Pandemic Is Intricately Linked to Biodiversity Loss and Ecosystem Health." *The Lancet Planetary Health*, vol. 5, no. 11, 2021, pp. e840–e850.

Le Guin, Ursula. *Always Coming Home*. London: Gollancz, 1985.

Levitas, Ruth. *The Concept of Utopia*. Oxford: Peter Lang, 1990.

Lewis, Tyson E., and Richard Kahn. *Education out of Bounds: Reimagining Cultural Studies for a Posthuman Age*. New York: Palgrave MacMillan, 2010.

Lorde, Audre. *The Master's Tools Will Never Dismantle the Master's House*. London: Penguin Books, 2018.

Lynch, Rebecca, Tristan Sturm, and Joseph Webster. "Apocalyptic Futures: Morality, Health and Wellbeing at the End of the World." *Anthropology & Medicine*, vol. 28, no. 1, 2021, pp. 1–12.

Marani, Marco, et al. "Intensity and Frequency of Extreme Novel Epidemics." *PNAS*, vol. 118, no. 35, 23 August 2021, p. E2105482118, https://www.pnas.org/doi/10.1073/pnas.2105482118.

Marks, Peter. *Imagining Surveillance: Eutopian and Dystopian Literature and Film*. Edinburgh: Edinburgh University Press, 2015.

McKay, George. *Senseless Acts of Beauty*. London: Verso, 1996.

McIntosh, Alastair. "Foreword." *Future Ethics: Climate Change and Apocalyptic Imagination*, edited by Stefan Skrimshire. London: Continuum, 2010, pp. vii–xi.

Minteer, Ben A. "The Perils of De-extinction." *Minding Nature*, vol. 8, no. 1, 2015, pp. 11–17.

Mohr, Dunja. "Critical Hope: Relationalities in 21st-century Speculative Fiction and Art." *Intersectional, Feminist, and Non-Binary Approaches in 21st Century Speculative Culture*, edited by Tomasz Fisiak, and Katarzyna Ostalska. London: Routledge, 2022, pp. 61–77.

Moore, Jason W. *Capitalism in the Web of Life: Ecology and the Accumulation of Capital*. London: Verso, 2015.

———. "The Capitalocene Part I: On the Nature and Origins of Our Ecological Crisis." *The Journal of Peasant Studies*, vol. 44, no. 3, 2017, pp. 594–630.

More, Sir Thomas. *Utopia*. Unabridged Edition. 1516. New York: Dover Thrift Editions, 1997.

Morton, Timothy. "Ecology without the Present." *Oxford Literary Review*, vol. 34, no. 2, 2012, pp. 229–38. *JSTOR*, www.jstor.org/stable/44030884.

———. *Hyperobjects: Philosophy and Ecology after the End of the World*. Minneapolis: University of Minnesota Press, 2013.

Mould, Oli, et al. "Solidarity, Not Charity: Learning the Lessons of the COVID-19 Pandemic to Reconceptualise the Radicality of Mutual Aid." *Transactions of the Institute of British Geographers*, vol. 47, no. 4, 2022, pp. 866–79.

Moylan, Tom. *Becoming Utopian: The Culture and Politics of Radical Transformation*. London: Bloomsbury Academic, 2021.

Moylan, Tom, and Raffaella Baccolini. "Introduction: Utopia as Method." *Utopia, Method, Vision: The Use Value of Social Dreaming*, edited by Tom Moylan and Raffaella Baccolini. Bern: Peter Lang, 2007, pp. 13–24.

Murphy, Patrick D. "Introduction." *Dystopias and Utopias on Earth and Beyond: Feminist Ecocriticism of Science Fiction*, edited by Douglas A. Vakoch. Oxford: Taylor & Francis, 2021, pp. 1–7.

Nash, Jennifer C. "Practicing Love: Black Feminism, Love-Politics, and Post-Intersectionality." *Meridians*, vol. 19, no. S1, 2020, pp. 439–462.

Nietzsche, Friedrich. *Human, All Too Human: A Book for Free Spirits*. 1878. Translated by R J. Hollingdale. Cambridge: Cambridge University Press, 1996.

Niranjan, Ajit. "'Era of global boiling has arrived,' says UN Chief as July Set to be Hottest Month on Record." *The Guardian*, 27 July 2023, https://www.theguardian.com/science/2023/jul/27/scientists-july-world-hottest-month-record-climate-temperatures

Nussbaum, Martha. *Frontiers of Justice*. Cambridge, MA: Harvard UP, 2006.

O'Connor, Brian, et al. "Monitoring Global Changes in Biodiversity and Climate Essential as Ecological Crisis Intensifies." *Ecological Informatics*, no. 55, 2020, p. 101033. https://doi.org/10.1016/j.ecoinf.2019.101033.

Pohl, Nicole. "Utopia Matters." *Spaces of Utopia: An Electronic Journal*, no. 7, 2009, pp. 1–14.

Popper, Karl. *The Open Society and Its Enemies*. 1945. London and New York: Routledge, 2002.

Preston, John, and Rhiannon Firth. *Coronavirus, Class and Mutual Aid in the United Kingdom*. London: Palgrave, 2020.

Puar, Jasbir. "Queer Times, Queer Assemblages." *Social Text*, vol. 23, nos. 3–4, 2005, pp. 121–39. https://doi.org/10.1215/01642472-23-3-4_84-85-121.

———. *Terrorist Assemblages: Homonationalism in Queer Times*. Durham, NC: Duke University Press, 2007.

Quammen, David. "Planet of Weeds." *Natural Acts: A Sidelong View of Science and Nature*. New York: Norton, 2009, pp. 161–88.

———. *Spillover: Animal Infections and the Next Human Pandemic*. New York: Norton, 2012.

Raekstad, Paul, and Sofa Saio Gradin. *Prefigurative Politics: Building Tomorrow Today*. Cambridge: Polity Press, 2020.

Richardson, Katherine, et al. "Earth Beyond Six of Nine Planetary Boundaries." *Science Advances*, vol. 9, no. 37, 13 September 2023. https://doi.org/10.1126/sciadv.adh2458.

Robinson, Kim Stanley. "The Coronavirus Is Rewriting Our Imaginations." *The New Yorker*, 1 May 2020. https://www.newyorker.com/culture/annals-of-inquiry/the-coronavirus-and-our-future.

———. "Keynote Address II–Kim Stanley Robinson–Living in the End Times." *Living in the End Times: Utopian and Dystopian Representations of Pandemics in Fiction, Film and Culture* conference, Cappadocia University, Turkey, 13 January 2021. Kapadokya Üniversitesi, *Youtube*, 13 Jan. 2021, youtube.com/watch?v=saVkaueMsuQ.

Robinson, Nick. "Economy: There Is No Alternative (TINA) Is Back." *BBC News*, 17 December 2023. https://www.bbc.co.uk/news/uk-politics-21703018.

Rose Bird, Deborah. "Shimmer: When All You Love Is Being Trashed." *Arts of Living on a Damaged Planet: Ghosts and Monsters of the Anthropocene*, edited by Anna Lowenhaupt Tsing, Nils Bubandt, Elaine Gan, and Heather Anne Swanson. Minneapolis: University of Minnesota Press, 2017, pp. G51–G63.

Sargent, L. T. "The Three Faces of Utopianism Revisited." *Utopian Studies*, vol. 5, no. 1, 1994, pp. 1–37.

Sargisson, Lucy. *Utopian Bodies and the Politics of Transgression*. London: Routledge, 2002.

Schmeller, Dirk S., Franck Courchamp, and Gerry Killeen. "Biodiversity Loss, Emerging Pathogens and Human Health Risks." *Biodiversity and Conservation*, vol. 29, 2020, pp. 3095–102.

Scott, Conrad. *Here, at the End: Contemporary North American Ecocritical Dystopian Fiction*. 2019, University of Alberta, PhD dissertation. https://era.library.ualberta.ca/items/1c6e0f88-52f4-43e6-a4d1-73dc6264577e.

Stirner, Max. *The Ego and Its Own*. 1844. Trans. Steven Byington. London: Rebel Press, 1993.

Szerszynski, Bronislaw. "The End of the End of Nature: The Anthropocene and the Fate of the Human." *Oxford Literary Review*, vol. 34, no. 2, 2012, pp. 165–84. *JSTOR*, www.jstor.org/stable/44030881.

Taylor, Taryne Jade, Isaiah Lavender III, Grace L. Dillon, and Bodhisattva Chattopadhyay. *The Routledge Handbook of CoFuturisms*. New York and London: Routledge, 2024.

Thomas, Leah. *The Intersectional Environmentalist: How to Dismantle Systems of Oppression to Protect People + Planet*. London: Souvenir Press, 2022.

Tsing, Anna Lowenhaupt. *The Mushroom at the End of the World: On the Possibility of Life in Capitalist Ruins*. Princeton, NJ: Princeton University Press, 2015.

Tsing, Anna Lowenhaupt, Nils Bubandt, Elaine Gan, and Heather Anne Swanson, editors. *Arts of Living on a Dying Planet: Ghosts and Monsters of the Anthropocene*. Minneapolis: University of Minnesota Press, 2017.

Van Dooren, Thom. *Flight Ways: Life and Loss at the Edge of Extinction*. New York: Columbia University Press, 2014.

Van Dooren, Thom, and Deborah Bird Rose. "Storied-Places in a Multispecies City." *Humanimalia*, vol. 3, no. 2, 2012, pp. 1–27.

WWF. *Living Planet Report 2022: Building a Nature Positive Society*. Edited by Rosamunde E.A. Almond, Monique Grooten, Diego Juffe Bignoli, and Tanya Petersen. WWF, Gland, Switzerland.

Yeomans, Sarah. "Pandemics in Perspective." *Bible History Daily*, Biblical Archeology Society, 26 August 2020. https://www.biblicalarchaeology.org/daily/pandemics-in-perspective/.

Zalasiewicz, Jan. "The Extraordinary Strata of the Anthropocene." *Environmental Humanities: Voices from the Anthropocene*, edited by Serpil Oppermann and Serenella Iovino. London: Rowman and Littlefield, 2017, pp. 29–45.

Zalasiewicz, Jan, et al. "Stratigraphy of the Anthropocene." *Philosophical Transactions of the Royal Society A*, vol. 369, 2011, pp. 1036–55. https://doi.org/10.1098/rsta.2010.0315.

Zizek, Slavoj. *Living in the End Times*. London: Verso Books, 2011.

Part 1

Monsters and Monstrosity

1 "In the woods the Tox is still wild"

The EcoGothic in Rory Power's *Wilder Girls*

Tânia Cerqueira[1]

Introduction

Young people have shown concern about the environmental futures humanity might face due to climate change, ranging from species extinction, toxic air pollution, rising sea surface temperatures, and sea-level rise to intense droughts, floods, and the globalized spread of infectious diseases. They show their unease through calls for climate action, and imagining their own futures, often drawing on speculative fiction (Priyadharshini). Young adult (YA) speculative fiction becomes a staple of these imagined futures, fueling the imagination while actively engaging with climate change and its repercussions. A liminal body of literature between children's and adult fiction, YA novels tackle environmental emergencies by moving away from oversimplified and depoliticized ecological crises easily solved in children's fiction (Echterling 286), and featuring a hopeful optimism, unlike much adult fiction. As Lindsay Burton puts it, the adolescent turmoil presents "a potent literary representation of authentic hope in the face of equally authentic destruction that has much to teach us about Anthropocentric presents and futures" (30). Because of its liminality, YA fiction becomes indispensable in approaching Anthropocentric environmental catastrophe.

In a powerful way, YA fiction brings into the open the transformative impact of climate change on the planet through the construction of environments profoundly altered by its effects. For instance, in Sherry L. Smith's *Orleans* (2013), the Gulf Coast is destroyed by a string of devastating hurricanes and a blood-borne virus called Delta Fever. In Neal Shusterman's and Jarrod Shusterman's *Dry* (2018), a lengthy drought in California leads to closing the water supply and friends and neighbors, crazed by thirst, turn on each other. In Darcie Little Badger's *A Snake Falls to Earth* (2021), the Dallas toad species faces extinction due to human environmental destruction. The environments in these novels are often menacing and harmful, and the adolescent protagonists must adapt to and/or enmesh with the natural world in order to grow and survive. Environmental crisis in YA fiction, Alice Curry argues, causes a crisis of embodiment, "reflected in the contentious relationships between the young protagonists of the novels and their social and ecological surroundings, relationships that are enacted on the discursive site of their own bodies" (15–16). Their bodies are thoroughly connected with the harmful, decaying

DOI: 10.4324/9781003345770-3

environment, frequently leading to transgressions of bodily boundaries. The teenage body (and self) and the environment intertwine, constructing new forms of being. Rory Power's *Wilder Girls* (2019) represents precisely this entanglement between body and nature.

In the novel, humanity's out-of-control actions against nature create an unknown disease, the Tox, which takes hold of Raxter, an isolated island off the coast of Maine, home to the Raxter School for Girls, a private school. The Tox has its roots in a prehistoric worm. Freed from its home in the Arctic ice due to the rising temperatures, it takes hold of every living being on Raxter (Power 341–42). The students are afflicted by the disease, which manifests itself in flare-ups with deadly consequences: most girls perish from the sickness, or from the deadly violence it provokes that incites them to kill each other. Those who survive the flare-ups develop bodily mutations: gills; a taloned hand; two heartbeats; a third, closed eyelid; and a serrated ridge of bone down the back. In *Wilder Girls*, anthropogenic climate change threatens the environment, bringing forth an infection that mutates everything it touches and turns nature into a nightmarish site.

In the emergent critical field of the ecoGothic,[2] nature is represented "as a space of crisis [that] conceptually creates a point of contact with the ecological" (Smith and Hughes 3). EcoGothic texts employ familiar Gothic tropes within an environmental context to discuss climate change and ecological destruction through the construction of an uncanny natural world. In the dystopian world of *Wilder Girls*, nature, transformed by the Tox, is described as monstrous and predatory, threatening the girls' survival; in the novel, nature is represented as a Gothic site. By critically engaging with the ecoGothic, I examine the intersections between the Gothic imagination and the natural world in *Wilder Girls*, emphasizing the island's woods. Utterly changed by the Tox, they have become savage and dangerous. Most importantly, I argue that the connection between the Gothic mode and nature intertwines with environmental awareness. I accordingly look into the theoretical considerations of ecoGothic, and analyze the intersection between Gothic tropes and the woods. Since the novel questions the human/nature binary through disrupting boundaries and blurring the lines of what it means to be human, I observe the relationship between the ecoGothic and posthumanism. As I explore *Wilder Girls* through these theoretical frameworks that move away from an anthropocentric worldview, I contend that the novel calls for a turn to ecocentrism.

Even though *Wilder Girls* is marketed as a YA dystopian book with horror-related elements, I chose to explore the text through the field of ecoGothic, a theoretical lens, instead of ecohorror, a genre. My choice falls to the ecoGothic due to Elizabeth Parker's considerations in *The Forest and the EcoGothic: The Deep Dark Woods in the Popular Imagination* (2020), where she dwells on the two concepts' differences. Parker argues that, in the ecoGothic, there is centrally "an undeniable sense of Gothic *ambience* in the natural environments" (*Forest* 35). In *Wilder Girls*, the forest takes front stage and is an entity described as dark, eerie, mysterious, and looming; that is, a Gothic atmosphere cloaks the forest in the novel. Additionally, the suffixes "horror" and "Gothic" have different connotations. Horror, as Parker justifies, implies "bad": thus, "in eco-horror, Nature will be somehow monstrous and dangerous to humankind" (36). Though the Gothic evokes fear, it also relates to desire, allowing for the exploration of the intricate human/nonhuman relationship. Hence, "Nature is something precious, to be protected—but it is also something terrifying, to be destroyed" (36). A central theme in *Wilder Girls*, because of the girls' bodily transformations, is the human/nonhuman relationship. Although ecohorror also takes into consideration this relationship, as Christie Tidwell claims, since the blurring of lines takes place within the horror genre, ecohorror narratives commonly ask "to consider the dangers of such interconnectedness" (117), making the human/nonhuman relationship something to fear. In contrast, the ecoGothic permits looking into the human/nonhuman relationship under a positive light and through ecocentric views.[3]

Fearing Nature: Approaching the EcoGothic

The study of texts about the relationship between humanity and nature, whether overtly Gothic or not, reveals that ecocritical concerns have always existed, haunting readers with "anxieties about the dangers of the natural world and our place in it, ethical perils of unchecked scientific experimentation and extractive industries, the unstable boundary between human and nonhuman, and a growing dread over human-caused environmental change" (Hillard 30). Home to anxieties and hauntings, the Gothic becomes the ideal mode to explore our horrors and concerns about nature and ecological devastation. The emergence of the ecoGothic theoretical lens is an expected development.

Simon C. Estok sowed the seeds for the development of the ecoGothic by introducing the concept "ecophobia," "an irrational and groundless hatred of the natural world" ("Theorizing in a Space" 208). In other words, ecophobia is humans' fear of the natural world—a fear rooted in the belief that humanity stands at the center of the world and nonhumans are to be (ab)used (Estok, "Theorizing in a Space" 216–17). Ecophobic narratives "are suffused with loathing, fear, disgust and horror, often attributing a capacity for retribution to a vengeful Nature, personified as malevolent antagonist" (Deckard 174). These elements are key to the Gothic. Thus, ecophobia is an integral part of ecoGothic texts. In fact, the ecoGothic emerges from this terror of nature.

In the introduction to *EcoGothic* (2013), Andrew Smith and William Hughes assert that this new critical field explores the Gothic mode through the lens of ecocriticism. Smith and Hughes contend that the ecoGothic is well-positioned to encapsulate climate change and environmental damage anxieties, becoming invaluable for cultural and political discussions about ecology and environmental awareness (5). David Del Principe claims the ecoGothic approach takes "a non-anthropocentric position to reconsider the role that the environment, species, and nonhumans play in the construction of monstrosity and fear" (1). That is, through evoking feelings of anxiety and fear, the ecoGothic aims to comprehend human estrangement from the nonhuman, and especially from nature. Sharae Deckard also stresses this relationship by arguing that the ecoGothic "represents the cultural anxieties about the human relationship to the nonhuman world through uncanny apparitions of monstrous nature" (174). She explains that the ecoGothic offers "a portal into the contemporary imagination of compound ecological crisis with complex temporal antecedents, to materialise with an uncanny immediacy those revenants of 'undead' processes in the past that continue to shape contemporary environments" (Deckard 186). The return of the undead, haunting past is one of the most pervasive Gothic themes. The ecoGothic, too, forcefully unearths the past, exposing how it shapes today's environmental crisis. Estok argues that the ecoGothic theorizes about menace, monstrosity, humanity's control-freak behavior, ontological and existential matters, and solutions ("Theorising the EcoGothic" 34). Through theorizing about solutions for environmental crises, the ecoGothic stresses a shift from anthropocentric to ecocentric views, driving humanity to look at the roots of environmental concerns. Parker suggests that "the ecoGothic is a flavoured mode through which we can examine our darker, more complicated cultural representations of the nonhuman world—which are all the more relevant in times of ecological crisis" (*Forest* 36). The nonhuman and their world(s) are central to the ecoGothic, which is concerned with ecocentric texts and views that shed light on ecophobic anxieties and fears, problematizing how those views help catalyze climate crisis.

The ecoGothic is, then, a theoretical lens that questions the relationship between the human being and the natural world—a relationship characterized by humanity's fear of losing control over nature[4] and of its agency, which turns nature into a place of monstrosity and terror that must be tamed. In *Wilder Girls*, nature, transformed by the Tox, flourishes wildly, reclaiming the space that was once its own—a reclamation that portrays Raxter's nature as a Gothic setting. Indeed, the ecoGothic employs familiar Gothic tropes and motifs, such as isolation, disease, madness, darkness, fear, and monstrosity within an environmental context. These haunting Gothic tropes manifest in the body of nature, leading to ecoGothic environments being commonly described as excessive: these are "sites of monstrous fecundity that threaten human civilization, where vines and vegetation run rampant, or where plagues and vermin spread deliriums" (Deckard 174). In creating an eerie and excessive nature, *Wilder Girls* confronts readers with ecoGothic views. In the face of nature's agency, the characters perceive the woods as monstrous and a site of seclusion, uncanniness, and disease.

"It's so quiet that you can hear the woods": The Gothic Forest

Hetty, the novel's main character, describes Raxter's nature as untamed, a place of shadows that is threatening and forbidding and permeated by disease, darkness, and fear. She observes that nature has become appalling and predaceous. Trees are crawling closer to the fence that separates the school from the forest. Their branches reach over the fence, "loaded with berries the color of blood [that when broken] open, [reveal] insides … black and oozing" (Power 58). Pines are taller and broader than they should be. Their canopies filter the sun and turn the little light that creeps into the forest phantasmagorical. In the forest, Hetty says, everything "feels forgotten, like we're the first people here in a hundred years.

No tire tracks left on the road, no sign this was ever anything but what it is now. We shouldn't be here. This place isn't ours anymore" (Power 59). Hetty's thoughts mirror humanity's belief of owning nature, as well as a fear of nature's agency.

In *Wilder Girls*, the natural world and the Tox are given agency by being anthropomorphized. For instance, when it is quiet, Hetty explains, she can hear the woods "growing and moving" (Power 60). The Tox inhabits the woods, where it "is still wild … blossom[ing] and spread[ing] with a kind of joy." In the woods, the Tox is "[u]nbridled and vicious and free" (Power 50–51). This anthropomorphized depiction of the woods and the Tox, which brings forth ecophobic fears, exacerbates the sense of terror the forest already produces as a Gothic setting.

Parker affirms that the forest, as a site of trial, trepidation, and terror, "has always been a mainstay of the Gothic" ("EcoGothic Secrets" 23). In a later study, Parker explains that the forest is feared because it is against civilization, is associated with the past, is a landscape of trial, is a setting in which we are lost, is a consuming threat, is a site of the human unconscious, and is an anti-Christian space. To her, "[t]he Gothic forest, in order to *be* Gothic, must always carry the discernible threat of at least one of these seven reasons" (*Forest* 48). The forest in *Wilder Girls* is associated with the past, is a consuming threat, and is a site of the human unconscious, as observed in the following analysis.

The past haunts Raxter, whether it is in the bleak classrooms where once the girls learned Math, French, and Physics, or in the girls' scars that tell of their traumatic survival of the Tox. The parasite itself is a memento of a haunting past. Indeed, Sladja Blazan discusses the notion of haunting and its relationship to nature. According to Blazan, "[a]s an active engagement with the past in the present, the classic gothic trope of haunting inevitably excavates histories within Nature" (4). To her, haunting "marks an affirmation, as spaces (including bodies) are haunted by something that has once existed in the past and returns insistently in the present …. Haunting marks bodies (including landscapes) as interactive spaces where death does not mean decay" and, for this reason, "our current ghosts, come to the fore the strongest in so-called natural environments, as these were often perceived as dead and now seem to come alive repeatedly in what is presented as revengeful and frightening ways" (5). The haunting past is entrenched in the land, connecting past and present through histories and stories brought back by an active nature. *Wilder Girls* represents this hauntingly connection through its climate change induced disease. Humanity's actions against nature, believed to have faded, bring a dormant creature from a prehistoric past into the present. The parasite, a ghost, returns to haunt the human being, who has ignored the effects of anthropocentric action for far too long. What was thought to be dead and buried is alive and has reached out from the depths of Earth to haunt the girls on Raxter, including through its effects on the forest.

Parker clarifies that the forest is a consuming threat because it "is a setting in which we fear being *eaten*: be it by literal predators such as wolves and bears, or by the many monsters that we imagine within it" (*Forest* 54). On Raxter, the threat of being consumed by the monsters of the forest is no longer an imagined menace but rather a reality. The Tox has also transformed Raxter's four-legged inhabitants, who now prey and feast on the girls. Furthermore, as Hetty puts it, the prehistoric worm lives "in our bodies, making us their own" (Power 340). The parasite feeds on both animal and human hosts, and, as it is later revealed, the host needs the parasite to survive, having both developed a symbiotic relationship.

Finally, the idea of the forest as a site of the human unconscious intertwines with the concept of the *unheimlich*: Raxter's forest was once familiar but has now become disturbing and strange. Although the woods look the same, the Tox made them strange and cruel. Everywhere Hetty looks, "there's something to be scared of" (Power 193). Sigmund Freud theorized the *unheimlich* as "that species of frightening that goes back to what was once well-known and had long been familiar" (124). This feeling usually arises when someone is confronted with "something that was long familiar to the psyche and was estranged from it only through being repressed" (148). Freud's word echoes F. W. J. Schelling's conceptualization of the *unheimlich*, where "something that should have remained hidden and has come in the open" (qtd. in Freud 148). In Power's novel, the resemblance between the parasitic pandemic and the return of the repressed, a popular Gothic theme that unveils secrets and social truths one wishes to keep hidden, is undeniable. The parasite came into the open to expose humanity's responsibility for the disease. After all, the Tox is nothing more than the result of humanity's extractive and capitalist behavior towards nature.

The premise of diseases and/or pandemics emerging due to humanity's action against the environment is not unusual in YA speculative fiction. Besides *Wilder Girls*, among other examples, Smith's *Orleans* and Cherie Dimaline's *The Marrow Thieves* (2017) take on this premise. In Smith's *Orleans*, which uses Hurricane Katrina as a framework, Delta Fever is a consequence of humanity's ill-treatment of nature. Katrina itself is a manifestation of humanity's environmental actions. Destructive hurricanes such as Katrina, which are becoming more common and violent due to rising global temperatures,[5] are non-natural catastrophes caused by humanity's actions over nature. Hence, as Roberta Selinger Trites proposes, in *Orleans*, "the Fever would not exist had manmade conditions not allowed it to spiral out of control as a pathogen" (109). Likewise, in Dimaline's *The Marrow Thieves*, when water sources start drying up, the United States of America builds pipes on Indigenous lands to steal water resources—actions that pollute the Great Lakes. Due to climate change, the world's edges are "clipped by the rising waters, tectonic shifts, and constant rains. Half the population [i]s lost in the disaster and from the disease that spread[s] from too many corpses and not enough graves" and those who are left stop "reproducing without the doctors, and worst of all, they [stop] dreaming" (Dimaline, "Story: Part One"). This dreamlessness results in a plague of madness. Non-Indigenous people who no longer dream lose their minds, killing themselves and others, and Indigenous people are hunted for their bone marrow (which in the novel houses the ability to dream). This widespread madness is also rooted in humankind's exploitative behavior towards nature—a behavior that changes not only nature, but humans themselves.

In *Wilder Girls*, besides the forest, the girls' bodies are also represented as a site of uncanny feelings. For example, Hetty feels unsettled when she sees Taylor's naked torso. On Taylor's chest is a cord of muscle the width of a thumb. It is "pale blue and twisting, almost braided, with a pulse to it like it's alive" (Power 288). The once familiar body has been de-familiarized by the Tox. Looking at themselves and the others, the girls experience the horror of the *unheimlich*, "an experience of otherness within sameness" (Beal 5). They experience the restlessness and fright that arises when something well-known suddenly becomes unfamiliar and petrifying. Consequently, their bodies are embedded with uncanny power—a power that comes from their bodies being posthuman.

The Posthuman Body: Destroying Human–Nature Boundaries

The posthuman body is a core theme of posthumanism.[6] Pramod K. Nayar divides posthumanism into transhumanism and critical posthumanism. The latter, he clarifies, "does not see the human as the centre of all things: it sees the human as an instantiation of a network of connections, exchanges, linkages and crossing with all forms of life" (5). This philosophical perspective decenters the human subject, exploring the close relationship of the human being with the surrounding world and challenging hierarchal binaries. It refuses humanism's views of man as the center of the universe and its exclusionary practices where "we can find the origins of sexism, racism and other exclusionary practices" (9). In this sense, posthumanism denies fallacious stories that have legitimized humanity's action over nature; it refuses

> [t]he stories we have been telling ourselves about human exceptionalism (we're the image of God), human entitlement (we're masters of this planet), and human identity (we're separate from and above "nature") [which] channeled our creativity into projects that transformed the planet—in our eyes—into a purely human domain. (Oziewicz et al. 1)

Posthumanism criticizes anthropocentrism and speciesism. Thus, posthumanism and the ecoGothic share the same views regarding the human/nonhuman relationship, repudiating humanist understandings. As Del Principe underlines, "the EcoGothic examines the construction of the Gothic body—unhuman, nonhuman, transhuman, posthuman, or hybrid—through a more inclusive lens, asking how it can be more meaningfully understood as a site of articulation for environmental and species identity" (1). Analogous to posthumanism, the ecoGothic shakes beliefs of what might be considered "human" and questions the human species' identity through the critique of oppressive hierarchal relationships. It destabilizes the human/nature binary in environmental spaces that draw together ecocriticism with humanity's more terrifying notions on nature.

In *Wilder Girls*, the Tox transforms the body, leading to mutations that cause the girls to inhabit posthuman bodies. The physical differences not only make the girls posthuman but also trans-corporeal. A living embodiment of Stacy Alaimo's trans-corporeality, defined as a movement across bodies and nature that alters the sense of self ("Trans-Corporeality" 435), the girls stand for humanity's ecophobic relationship to nature and must be contained. Unable to control the Tox, nature, or the girls, the Centers for Disease Control and Prevention (CDC), which has been on Raxter studying the Tox for over a year, abandons the island and orders its bombing, hoping to erase evidence of their presence and their testing on the girls, and to prevent an outbreak. In an interview, Power addresses the bodily transformations her characters experience: "The actual changes they undergo are each based on a physical adaptation that exists in the natural world …. Taylor's adaptation … is based on something called a lateral line, which fish use to sense movement around them" (Power, "Editor/Author Interview"). As Byatt, Hetty's closest friend, realizes, the Tox "models us after the animals around us, tries to change our bodies, push them further than they're willing to go. Like it's trying to make us better, if only we could adapt" (Power 169). The Tox metamorphoses them to survive in harsh environments instigated by anthropocentric climate change. Unfortunately, their posthuman bodies threaten humanity's views about itself as they expose how deeply interconnected the human body is with nature.

The character that is most affected by the Tox is Mr. Harker, Raxter's caretaker. Due to having lived in the woods for over a year, he intertwines with the surrounding landscape in ways that none of the girls ever do: his skin is bleached and pulled, his mouth sprouts roots, and he has "[b]ranches burrowing in ears and under fingernails and slinking down his arms" (Power 212). Besides the roots, vines, and branches sprouting from him, Mr. Harker's body houses mold and wingback beetles. His rib cage is made of "branches, splendid ribs curving and cresting," and his heart is described as "[b]uilt from the earth, from the bristle of pine" (215). When she sees the man-shaped figure, Hetty identifies him as Mr. Harker "[u]ntil he eases into the red glow of the flare light, and it isn't anymore" (211). When Hetty sees his body clearly, Mr. Harker is no longer described by the pronoun "he" but by "it." To Hetty, his non-normative human body takes his humanity from him. In the fight that ensues between them, Hetty kills Mr. Harker. However, the reader must wonder if Mr. Harker truly meant to attack Hetty and Reese, his daughter, or if he was defending himself (Hetty points a shotgun at him) or was communicating the only way he knew—through rattling sounds and writhing vines. Although Hetty herself inhabits a posthuman body, her conceptualization of Mr. Harker as nonhuman—as belonging to other species—seals his fate. After all, while she only develops a third eyelid that permanently closes one of her eyes, Mr. Harker's whole body changes. This instance reveals the deeply rooted speciesism that permeates Western culture and discourses. Speciesism, Nayar explains, "positions the human as the dominant species that then controls, domesticates, oppresses, exploits, guards and pets non-human, animal species" (96). Humanist, anthropocentric thought has built a violent discourse against the nonhuman. Humanity exploits nature as it sees fit to maintain its exceptionalism and privilege. Nevertheless, unfortunately, "the discourse and practice of speciesism in the name of liberal humanism have historically been turned on other *humans* as well" (Wolfe 37). Unsurprisingly, the teenage girls, who are no longer recognized as human, are cruelly treated by those who should help them and are left to die on the island.

Ultimately, *Wilder Girls* undermines anthropocentric discourses by highlighting the girls' plight and the humanity within their posthuman bodies. As readers are confronted with the girls' bodies that blur the boundary between humanity and nature, they are prompted to ask whether it is acceptable to commit acts of violence against nature and other beings because of their bodily difference or species. The novel rejects the human/nature binary, challenging and ripping to pieces the belief that the human owns and stands apart from the nonhuman (even if, at moments, the main character indulges in these views). By representing "human corporeality as trans-corporeality, in which the human is always intermeshed with the more-than-human world" (Alaimo, *Bodily Natures* 2), the novel unveils the irrefutable connection between humanity and nature.

Its ending[7] might seem gloomy, with Hetty, Reese, and a catatonic Byatt[8] adrift after escaping the island, their future uncertain. Nevertheless, there is hope in Hetty's thoughts as they sail away: "I don't know where we're going. I don't know what's next. But Reese's heartbeat is steady in my ear, and I remember—I remember how it was. The three of us together, and I'll make it that way again" (Power 353). Their survival does not allow the CDC to erase the impact of anthropocentric climate change on Raxter nor cover their mistreatment of nature and those inhabiting posthuman bodies. Hetty, Reese, and Byatt are living proof of the Tox and of the CDC's mistreatment. Through the non-annihilation of the Tox—it still lives within Hetty and Reese—and of every Raxter resident, the ending offers hope. On the one hand, the truth behind the Tox and the consequences of the rising of Earth's temperature are not expunged. On the other hand, the open ending allows readers to imagine possibilities—to wonder what can be done to not fall prey to a humanist, anthropocentric worldview that has had terrible consequences on Earth's ecosystems.

Conclusion: Moving towards Ecocentrism

The premise behind *Wilder Girls* is close to reality. In November 2022, news came out that scientists found a 48,500-year-old "zombie virus" buried in ice in Siberia (Birnbaum and Francis, "'Zombie' Viruses Are Thawing in Melting Permafrost"). These were followed by the publication of an exploration of bacteria and microorganisms trapped in the permafrost. As researchers explain, the planet's warming is causing the rapid permafrost thaw, resulting in the physical release of bacteria trapped in deep permafrost. They conclude by questioning the likeliness of viruses infecting a suitable host, and, even though it is impossible to find an answer, "the risk is bound to increase in the context of global warming, in which permafrost thawing will keep accelerating, and more people will populate the Arctic in the wake of industrial ventures" (Alempic et al. 12–13). Parallel to Power's novel, viruses and bacteria are becoming free from their icy confines, evincing what years of Western humanist action over nature has done to unbalance Earth's ecosystems.

EcoGothic texts demand "a serious shift in perspective from the pervasive Western anthropocentrism (a view of the world that is human-centric) to '*eco*centrism' (where all living things are of equal importance and nature is no longer defined in terms of human value)" (Parker, "'Just a Piece of Wood'" 218). *Wilder Girls* calls for a turn to ecocentrism. Through the depiction of a hostile, consuming nature and an environmental disease that un-tames everything it touches, the novel compels readers to face the ecophobic origins of human beings' relationship with nature: an anthropocentric discourse rooted in an entitlement that this planet is theirs to do with as they please and that allows for the destruction of everything that cannot be controlled and/or identified as the Other. Once humanity unburies the fears and prejudices that led to a climatic, anthropogenic environmental crisis that, sooner than later, will reach a point of no return, human beings can start looking for solutions and attempt to undo the damage caused to nature.

In Power's narrative, the ecoGothic allows readers to understand erroneous assumptions, embrace nature as equal, and urges a nonhuman turn that may avoid some of the dreadful fallouts of climate change ingrained in one's imagination. By creating a point of contact with the ecological, the ecoGothic reconciles eco-anxieties. Indeed, in *Wilder Girls*, the girls find power within their entanglement with an ecoGothic nature. For instance, Byatt does not wish to be cured. She wants to keep her mutations because, as she explains: "I think I'd been looking for it all my life—a storm in my body to match the one in my head" (Power 179). She finds strength in her posthuman body. Moreover, the ecoGothic finds answers to our ecological crisis through the acknowledgement of "the root issues that have brought us to where we are today" (Estok, "Theorising the EcoGothic" 49). If humanity recognizes its mistakes and starts living in harmony with nature, without preconceived ideas of superiority or ownership, it can finally acknowledge the fragility of the ecosystem and the importance of protecting it.

Inhospitable representations of nature may intensify humans' fear and fuel deep-seated ideas about subjugating the environment. However, I argue that these depictions will leave an imprint on readers, who will focus on generating futures of human/nature interconnectedness. If not, humanity's destructive ways may raze a nurturing being—in the end, the prehistoric worm in *Wilder Girls* was trying to adapt the girls, giving them the tools to strive in a landscape transformed by an anthropocentric climate crisis.

Notes

1 This work was supported by National Funds through FCT—the Portuguese Foundation for Science and Technology (2021.04547.BD.).

2 There is still uncertainty surrounding this new field, as the definitions explored show. This uncertainty is identified in the word itself. There are several variations in which "eco" and "Gothic" are written in lowercase or uppercase. I use the variation "ecoGothic." I have chosen to capitalise the word "Gothic," since renowned Gothic scholars such as David Punter, Glennis Byron, and Catherine Spooner capitalize it when referring to Gothic literary and cultural works, aesthetics, or to the mode itself. In my view, ecoGothic is a mode—hence my decision to capitalise "Gothic."

3 Although ecohorror intentionally raises environmental awareness (Parker, *Forest* 35), it highlights anthropocentric perspectives because it focuses on nature's revenge against human beings and humankind's efforts to fight them back.

4 Estok expresses that "the prospect of a loss of control—the perceived threat to human agency by nonhuman nature—is at its core ecophobic" ("Theorising the EcoGothic" 46). I would add that losing control is also at the core of ecoGothic, since not being in control of the nonhuman originates predatory, frightful portrayals of nature.
5 Global warming is raising sea surface temperatures. Scientists argue that two factors "contribute to more intense hurricanes: ocean heat and water vapor" and that these "have increased over the past 20 years because of human activities, such as burning fossil fuels and deforestation," since these activities "have significantly raised the CO2 levels in the atmosphere" ("Hurricanes and Global Warming").
6 For more on posthumanism, see N. Katherine Hayles' *How We Became Posthuman* (1999), Cary Wolfe's *What Is Posthumanism?* (2010), and Rosi Braidotti's *The Posthuman* (2013). On posthumanism and children's and YA fiction see Victoria Flanagan's *Technology and Identity in Young Adult Fiction: The Posthuman Subject* (2014), Zoe Jacque's *Children's Literature and the Posthuman: Animal, Environment, Cyborg* (2015), editors Anita Tarr and Donna R. White's *Posthumanism in Young Adult Fiction: Finding Humanity in a Posthuman World* (2018), and Jennifer Harrison's *Posthumanist Readings in Dystopian Young Adult Fiction: Negotiating the Nature/Culture Divide* (2019).
7 The novel's end seemly favours egocentric humanism and neoliberal individualism since Hetty and Reese abandon their classmates when they escape the school.
8 Before evacuating the island, a CDC doctor forces Byatt to breathe a gas meant to kill the Tox and, consequently, to kill her. Because of the gas, Byatt senses the parasite inside her and removes it with a scalpel. By extracting the parasite, Byatt survives. However, she is left in a catatonic state with no sense of self.

Works Cited

Alaimo, Stacy. *Bodily Natures: Science, Environment, and the Material Self.* Indiana UP, 2010.

———. "Trans-Corporeality." *Posthuman Glossary*, edited by Rosi Braidotti and Maria Hlavajova, Bloomsbury Academic, 2018, pp. 435–38.

Alempic, Jean Marie et al. "An Update on Eukaryotic Viruses Revived from Ancient Permafrost." *Viruses*, vol. 15, no. 2, February 2023, pp. 1–16. *MDPI*, doi: 10.3390/v15020564.

Beal, Timothy. *Religion and Its Monsters*. Routledge, 2002.

Birnbaum, Michael, and Ellen Francis. "'Zombie' Viruses Are Thawing in Melting Permafrost Because of Climate Change." *The Washington Post*, 2 December 2022, https://www.washingtonpost.com/climate-environment/2022/12/02/zombie-virus-russia-permafrost-thaw/. Accessed 17 May 2023.

Blazan, Sladja. "Haunting and Nature: An Introduction." *Haunted Nature: Entanglements of the Human and the Nonhuman*, edited by Blazan, Palgrave Macmillan, 2021, pp. 1–19.

Burton, Lindsay. "Playing with the Trouble: Children and the Anthropocene in Nnedi Okorafor's Akata Witch Series." *Fantasy and Myth in the Anthropocene: Imagining Futures and Dreaming Hope in Literature and Media*, edited by Marek Oziewicz et al., Bloomsbury, 2022, pp. 28–38.

Curry, Alice. *Environmental Crisis in Young Adult Fiction: A Poetics of Earth.* Palgrave Macmillan, 2013.

Deckard, Sharae. "Ecogothic." *Twenty-First Century Gothic: An Edinburgh Companion*, edited by Maisha Wester and Xavier Aldana Reyes, Edinburgh UP, 2019, pp. 174–88.

Del Principe, David. "Introduction: The EcoGothic in the Long Nineteenth Century." *Gothic Studies*, vol. 16, no. 1, May 2014, pp. 1–8. *Edinburgh UP Journals*, doi: 10.7227/GS.16.1.1.

Dimaline, Cherie. *The Marrow Thieves*. E-book ed., Cormorant Books, 2017. Kobo.

Echterling, Clare. "How to Save the World and Other Lessons from Children's Environmental Literature." *Children's Literature in Education*, vol. 47, no. 4, December 2016, pp. 283–99. *SpringerLink*, doi: 10.1007/s10583-016-9290-6.

Estok, Simon C. "Theorising the EcoGothic." *Gothic Nature*, no. 1, 2019, pp. 34–53. https://gothicnaturejournal.com/wp-content/uploads/2019/09/Estok_34-53_Gothic-Nature-1_2019.pdf.

———. "Theorizing in a Space of Ambivalent Openness: Ecocriticism and *Ecophobia.*" *Interdisciplinary Studies in Literature and Environment*, vol. 16, no. 2, Spring 2009, pp. 203–25. *Oxford Academic*, doi: 10.1093/isle/isp010.

Hillard, Tom J. Gothic. "Nature Revisited: Reflections on the Gothic of Ecocriticism." *Gothic Nature*, no. 1, 2019, pp. 21–33. https://gothicnaturejournal.com/wp-content/uploads/2019/09/Hillard_21-33_Gothic-Nature-1_2019.pdf.

"Hurricanes and Global Warming." *Climate Policy Watcher*, 3 May 2023, https://www.climate-policy-watcher.org/ocean-circulation/hurricanes-and-global-warming.html. Accessed 27 May 2023.

Oziewicz, Marek, et al. "Introduction: The Choice We Have in the Stories We Tell…." *Fantasy and Myth in the Anthropocene: Imagining Futures and Dreaming Hope in Literature and Media*, edited by Marek Oziewicz et al., Bloomsbury, 2022, pp. 1–11.

Parker, Elizabeth. "EcoGothic Secrets: Venturing Beneath the Deep Dark Forest." *Beasts of the Forest: Denizens of the Dark Woods*, edited by Jon Hackett and Seán Harrington, John Libbey Publishing, 2019, pp. 23–35.

———. "'Just a Piece of Wood': Jan Švankmajer's Otesánek and the EcoGothic." *Plant Horror: Approaches to the Monstrous Vegetal in Fiction and Film*, edited by Dawn Keetley and Angela Tenga, Palgrave MacMillan, 2017, pp. 215–25.

———. *The Forest and the EcoGothic: The Deep Dark Woods in the Popular Imagination.* Palgrave Macmillan, 2020.

Power, Rory. *Wilder Girls*. Ember, 2020.

Pramod, Nayar K. *Posthumanism*. Polity, 2014.

Priyadharshini, Esther. *Pedagogies for the Post-Anthropocene: Lessons from Apocalypse, Revolution & Utopia*. Springer, 2021.

Sigmund, Freud. *The Uncanny*. Translated by David McLintock, Penguin Books, 2003.

Smith, Andrew, and William Hughes. "Introduction: Defining the EcoGothic." *Ecogothic*, edited by Smith and Hughes, Manchester UP, 2013, pp. 1–14.

Tidwell, Christy. "Ecohorror." *Posthuman Glossary*, edited by Rosi Braidotti and Maria Hlavajova, Bloomsbury Academic, 2018, pp. 115–17.

Trites, Roberta Selinger. *Twenty-First-Century Feminisms in Children's and Adolescent Literature*. UP of Mississippi, 2018.

Wolfe, Cary. *Animal Rites: American Culture, the Discourse of Species, and Posthumanist Theory*. U of Chicago P, 2003.

2 The Human/Un(human)

Monster, Ecophobia, and the Posthuman Horror(scape) in Dibakar Banerjee's "Monster," *Ghost Stories*

Ujjwal Khobra and Rashmi Gaur

Introduction: Un-taming the Un(human)/Monster

The chapter aims to examine the relationality between the Human and the Un(human)/Monster[1] as an entanglement of contagious monstrosity and the liminal "in-betweenness"[2] of *being* human and *becoming*[3] un(human) in Dibakar Banerjee's short film, "Monster," from the cinematic horror anthology, *Ghost Stories* (2020). While contextualizing animality, consumption, and cannibalism as interactive taxonomical markers of identity formation, "Monster" presents a storied horror(scape) where humans feed on other humans to ensure their survival. They undergo a physical transformation alongside developing a primitive instinct of preying on and consuming the other, thereby situating the underlying praxis: "if you are not one of us, then you are our food." As an anamorphic relationality,[4] the "monster" functions as a polymorphous assemblage in Banerjee's short film. Further, it explicates the superimposed nature of the monster/human binary as "monsters" are an integral part of our lives, imagination, culture, and interconnected socio-political ontologies.

As a gateway to the posthuman universe, the term "monster" signifies "an embodiment of a certain cultural moment—of a time, a feeling, and a place," as theorized by Jeffrey Jerome Cohen in *Monster Theory* (1996, 4). On the other hand, as posthuman subjects, "monsters" refuse to fit within one category or pedagogy; they therefore provide an opportunity for appropriation, interpellation, and interpretation to nurture alternate worldbuilding. Simon C. Estok details the "cannibal as the perfect monster" (*Cannibalism* 3). The "cannibal" rewrites the "acceptable and the repugnant" by reassessing the porous boundaries between nonliving/living, contagious/non-contagious, and animal/human simultaneously (Estok 3). Taking a cue from Estok's understanding of the "monster," the "monster" in the film contextualizes the need to reimagine the Human as a category and its embodied anthropocentric situatedness to ensure sustainability and ecological sustenance.

A fraction of this endeavor is to acknowledge the convergence of challenges informing our present realities in the form of global warming, the escalating loss of biodiversity, climate change, and violence induced by exclusionary politics, coupled with the current planetary crisis in the form of COVID-19 pandemic. This reiterates the precarious nature of our quintessential "contained" existence,[5] where certain more-than-human "agents"[6] have become more critical than ever. As Frank

DOI: 10.4324/9781003345770-4

Kermode states in *The Sense of an Ending* (1967), crises are "inescapably a central element in our endeavors towards making sense of our world" (94).

Building on Kermode's modern-day interpretation of "crises," the acknowledgment of our dense web of interconnectedness, global vulnerability, and multispecies situatedness inaugurates both a call for action and a dream to cultivate empathy for all. The abovementioned planetary crises have found significant representations in literary and visual cultures, widely contributing to posthuman scholarship, where "humanness" is constantly being negotiated with, identified with, and dissociated from. Storying crises allow us to critically rethink our multiple, conjoined apocalyptic planetary challenges, and assume ethical responsibility toward intensifying ecological collapse, ecophobia, inter-/intraspecies relationality, and embodied otherness as posthuman actualizations.

In addressing some of these multifarious contestations connecting more-than-human entanglements and embracing the creative ontology embodied by "being-with-many,"[7] Dibakar Banerjee's "Monster" formulates an affirmative engagement with the "now"[8] as an ode to the living and the dead—"a cannibalized world where monstrosity becomes the norm, foregrounding vulnerability, naturalization of death, and violence towards the decomposed, disposable, and the damned 'Other(s)'" (Braidotti, *The Posthuman* 15). In contextualizing "otherness," the chapter draws on Simon C. Estok's concept of "ecophobia" as a new sensibility that informs the ways in which we are becoming more-than-(just-)human.

Here, Banerjee chronicles a tale of survival and becoming more-than-human at the edge of an ongoing apocalypse where humans start consuming other humans. The three unnamed characters—the "Outsider" (played by Sukant Goel), "Survivor 1" (a six-year-old boy, the last survivor of Smalltown), and "Survivor 2" (an eight-year-old girl)—attempt to survive in an un(human)/monster-infested Smalltown. The "monster," played by Gulshan Deviah, was Survivor 2's father and the leader, named "Alpha," of Bigtown before transmuting into a four-legged, human-eating monster with an insatiable thirst for blood. As the populace of Smalltown did not surrender their land to Bigtown, Alpha began consuming them. He devoured Survivor 1's father and annihilated the entire population of Smalltown. To ensure their survival, some individuals cannibalized others and, thereafter, metamorphosed into "monsters."

Estok states in his work, "Cannibalism, Ecocriticism, and Portraying the Journey," that the act of cannibalizing the other evokes horror and fascination on many levels (2). Banerjee's "monster"—an entity that functions by the force of consumption and rejects all forms of coexistence and cohabitation—evokes the horror and fascination of embodied, naturalized otherness. It fetishizes human flesh and is characterized by a pathogenic sensibility, which, in turn, performs cannibalism to grow and fester, making Smalltown an imagined horror(scape) in Banerjee's film.

The (Un)thinkable: "make you into a nice ham sandwich!"[9]

What evokes fear? As a pluri-directional "affective belonging," the assessment of phobia/fear introduces a self-reflexive inquiry into the basic proponents of being (vulnerably) "Human" (Braidotti and Hlavjova 13). Fear is an ongoing, ever-consuming, convoluted dialogue between humans and their environments. According to Estok, ecophobia in particular exists on a "spectrum" that incorporates aversion, contempt, apathy, disregard, and callousness toward nature (and nonhuman others) (1). The "ecophobic condition" acts upon the evolutionary imperative of self-preservation, which may lead to speciesism (Estok 1). The collective planetary fears in the form of multifaceted vulnerabilities, the horror inflicted by the COVID-19 contagion, environmental degradation, degrading multi-species network, and a consistent disregard for our more-than-human entanglements, have mutated into an unsettling predicament, which can be substantiated as the "contrast between how much we know and how little we have acted on this knowledge" (Oziewicz 1).

As a literary extrapolation from this unexpected realization of confronting multiple crises, certain cautionary yet haunting visions of dead, decaying, and apocalyptic landscapes have found precedence in our films and literary culture. Furthermore, the ongoing critical transitions introduced by many interdisciplinary approaches, such as Karen Barad's "interactivity" and Serenella Iovino and Serpil Oppermann's pathbreaking work on "material ecocriticism," along with many other critical thinkers in the posthumanist discourse, provide a resounding reiteration of Ihab Hassan's remarks on posthumanism. The escalating significance of comprehending our contemporary posthuman condition while contextualizing numerous unprecedented human-made crises that we face today has made Hassan's remarks more critical than ever:

> At present, posthumanism may appear variously as a dubious neologism, the latest slogan, or simply another image of man's recurrent self-hate. Yet posthumanism may also hint at a potential in our culture, hint at a tendency struggling to become more than a trend ... We need to understand that five hundred years of humanism may be coming to an end, as humanism transforms itself into something that we must helplessly call posthumanism. (Hassan 843)

What appeared as mere speculation in Hassan's 1977 essay today formulates the discursive and epistemological shifts in the perception of the "Human" as a category under perpetual deconstruction. Accordingly, the monster/human entanglement in Dibakar Banerjee's "Monster" underlines a similar dismantling of the normative singularity associated with the Human and the human body as (its) Eurocentric ideal, "the measure of all things" (Braidotti, *The Posthuman* 21). Interestingly, such narratives can also embody "ecophobia."

As an active site of "ecophobia" and apocalyptic dread, "Smalltown" is an uncanny landscape where human bodies are forced to consume other human bodies. Notably, a congruency is established between the burned landscape of Smalltown and blood-soaked menageries. In a particular scene, Survivor 1 informs the Outsider that the remaining population of Smalltown resorted to cannibalism, devouring their offspring and former acquaintances for survival. This community, where cannibalism becomes a shared social metaphor, corresponds to the pervasive persecution of those who do not conform to the act of eating flesh, thereby reinstating oppositional binaries such as "us" versus "them." Here, the "monster" serves as a cartographic tool to decode the ecophobic condition marked by phobic articulations evoked by human/un(human) becomings. Such a reading elicits critical concerns: are we still "human" when we become flesh for the "monster"? Or is it the phobic response instilled by the "monster" that, fundamentally, defines our becoming-human?[10]

As a (dis)figuration where "all forms come undone," Banerjee's monster emerges as "a post-anthropocentric assemblage," a subject in continuous formation that rejects prescribed modes of subjectivities in both popular cinematic imaginaries and critical humanities (Deleuze and Guattari 13; Braidotti; Cohen; Nayar; Mazurov). An "ontological liminality" functioning as the "Harbinger of Category Crises," the "monster" figuration offers a wide range of interpretations and dialogues between the Human and the more-than-human others that "it" consumes (Heise-von der Lippe 2; Cohen 6). For instance, Banerjee's monster substitutes humanness by eating human flesh. For it, human flesh acts as a digestible—it literally internalizes human-ness, even if it happens to be its own kin. The monster represents an embedded-embodied cultural filament of "nature" that is pathologized or characterized by ecophobia. The lack of familiarity with our speculative and hybrid counterparts such as the "monster" perpetuate un-philic-ness toward more-than-human worlds.

According to Estok, "writing monstrosity is the narrativization of ecophobia, imagining unpredictable agency in nature that must be subject to human power and discipline" (*Ecophobia,* 124). In depicting such temperamental "anxieties"[11] stemming from Human-Un(human) indeterminacy, Nikita Mazurov classifies the "monster" as not just "an abstract conceptualization, but an instantaneous, enacted entanglement" (262–63). Mazurov explains that the monster/un(human) is

> a continuous, unstable project of both disassembly or ex-figuration and of unsanctioned coupling; concrete and relational, it is a practiced hybridity of form which eludes conceptual formalizations, existing as it does as a state of contestation and troubling—shifting, adjusting, and dissolving at whim. (262)

However, it is not solely the visual manifestation of the "monster" on screen that contributes to ecophobia; rather, the sinister ecology of Smalltown, for instance, offers a detailed depiction of the dread that consumes its sanguinary landscape. In the opening frame, the dark screen is patchy, with the loud snapping and chirping of grasshoppers echoing like mysterious music of apocalyptic doom. This is followed by a pastoral background covered with ash-gray smog from burned wood, houses, and machines, with wild grass and a scarecrow clothed as a human. The deep blue sky and the displeasing green of the field offer glimpses of gray, black, and cyan portraits, setting the mysterious tone of the film. The seemingly abandoned village comes to life when the viewer hears an unnamed man enter this claustrophobic frame. He is an outsider who is unaware of the rules of survival in Smalltown. The town is depicted as a disconnected, uninhabitable, gory land with no electricity, basic infrastructure, food, or means of communication with the outside world. It is a land of missing, dead, and monster(ous) people, a cinematic horror(scape).

In the second shot of the film, the Outsider attempts to navigate the dark space using the flashlight on his phone and witnesses a boy, Survivor 1, dressed in shabby clothes and a broken helmet in a corner. The six-year-old unnamed boy is the last survivor of Smalltown. He speaks in a hushed tone and remains static to avoid being detected by the monster and other elders of Smalltown who have transformed into cannibals. The blank expression and the terror in his eyes reaffirm the "factuality of ecophobia" (Estok, *Ecophobia* 2), as his acts of self-preservation are a consequence and an affective response to the phobia generated by the devouring monster, who consumed his parents in front of him. As the Outsider attempts to talk to Survivor 1, the latter says, "Turn off the light if you want to live. If you see anything, do not move. When I tell you to run, run. Run!" (*Ghost Stories* 1:14:10). Through this scene, Banerjee portrays an unprecedented loss of human agency, the inability to function without phobia, restricted mobility, and a maddening fear of being consumed.[12] Survivor 1's actions make the monster's metaphorical omnipresence visible on screen. However, there is no escape in Banerjee's apocalyptic imaginary.

As the narrative advances, the Outsider follows Survivor 1 to a hut where he meets Survivor 2, an unnamed, 8-year-old girl, who points a knife at him as she presumes that all elders eat human flesh and transmute into monsters. However, Survivor 1 explains that the Outsider can talk, has not eaten human flesh, and might have come to their rescue. Survivor 2 remains defensive as she is sure that the "monster" terrorizing Smalltown can change form and shape, like her father, Alpha. He was the first to cannibalize and transform into a hairy, four-legged creature, and later devoured Survivor 1's father and the entire population of Smalltown. The two survivors come from different places; however, they attempt to survive together in the haunted space swarming with monstrous beings. For the survivors, the transmuted elders confirm an ongoing erasure of the world that they once habited as a safe space.[13]

The idea of speech and sight as differentiating between the human and the un(human) is an intriguing trope employed by Banerjee. The first sign of the loss of "humanness" is the loss of language, memory, and sight in human beings who devour human flesh. Often portrayed through incessant howling and an unending thirst for human blood, the "monster" of the film is presented as "a provocative figuration [as a mode] of knowledge production," an uncanny hybrid combining "apocalyptic visions of extinction with…(no) euphoric scenarios of escape to new extraplanetary homes" (Braidotti, *Feminism* 211). The monster, in this context, characteristically functions as a viral contagion. It has an "all-pervasive presence" in the "abruptly defamiliarized" space of blood-stained Smalltown (Parui and Raj 1438).

Similar to the viral contagion, the "monster" re-affirms an augmented sense of vulnerable existence since death and extinction are not dictated by the human subject. The monster creates an ontological order of a "contagious condition"—where the consumption of human flesh by other humans becomes "contagious," creating a (dis)assembly of Survivor 1 and 2 on the one hand and the unrecognizable monstrous bodies on the other, a "fantastic fabulation,"[14] manifesting our phobia, horror, and possible "collective" future on screen (Braidotti, *Feminism* 213). To this speculation, Braidotti presents a thought-provoking image of the "monstrous" anomaly as a mode of counteraction in her latest work, *Posthuman Feminism* (2022):

> Monstrous anomalous entities embody ontological impropriety. Their appearance is a provocation that upsets the status quo and evokes anxiety, or rather a mixture of fascination and loathing. Their metamorphic powers are immense: they act as mirrors to reflect our sense of inadequacy or inner monstrosity. (218)

Building on Braidotti's theorization, Banerjee's "Monster" not only "act[s] as a mirror" but embodies a shared sense of "inner monstrosity," theatrically depicted through metamorphosed animal-like anatomy—enlarged canines with distinct nasal fossa, protruding pointed ears, ungulate hooves, clawed appendages for hunting, and brown fur covered with blood and morsels of human flesh. The film presents a conventional image of a monster. However, it complicates the human/other nonrelationality by alchemizing human beings into literal "monsters." Therefore, the "monstrosity" projected through heinous acts of self-consumption (referring to intraspecies relationality) has a human past and a relation to the world "it" consumes (Estok, *Cannibalism* 2). Braidotti labels such narratives as the "inhuman edge of the posthuman condition" (*Feminism* 124). Subsequent to Braidotti's theorizations, Francesca Ferrando's cautionary remarks meditate on the uncertain nature of mortality, as she argues, "We can die 'now'! not tomorrow, not today, Now," an experientiality registered by the COVID-19 pandemic as we situate the living, the contagious, and the dead in the global catastrophe (Ferrando 00:02:26).

Likewise, Banerjee's film also challenges the ontological universality associated with "Anthropos" in revealing systemic and philosophical inconsistencies in "being" Human rather than "becoming" one.[15] For instance, in a pivotal scene, Survivor 1 engages Survivor 2 in an argument, wherein Survivor 1 raises concerns about the lack of resources and basic amenities in Smalltown. Before transmuting into a literal "monster," Alpha assumed control of Smalltown and imposed new mandates to compel the residents to officially seek permission from Bigtown to own and inhabit any space. This established an immediate terrestrial dichotomy between Bigtown and Smalltown. Furthermore, Survivor 1 suggests that a state of mutual harmony could have been achieved if Bigtown had deterritorialized its basic infrastructure and embraced a more inclusive approach of "becoming" more than just "being." However, as it happens, the violence, urgency to survive, and the perishability of all human life dictates the very existence of the survivors and the Outsider.

A Palimpsest of Intertextuality: Deconstructing the Un(Human) That We Are

Arjun Appadurai, in his work "Dead Certainty: Ethnic Violence in the Era of Globalization" (1998), provides an account of a new "kind" of "ethnic violence," which involves "mutilation, cannibalism, rape, sexual abuse, and violence against civilian spaces and populations" (907; Braidotti, *The Posthuman* 123). Appadurai explicates "violence" as an epistemic discourse that proliferated into multiple courses of expressions and manifestations especially among "socially proximate persons" and therefore becomes "uncertain" in cases of ethnicity-based violence (908). Within a consideration of Banerjee's "Monster," a film created within the ethnically complex Indian context, "cannibalism" functions as an "event" of "bodily brutality perpetrated by ordinary persons against other persons with whom they may have—or could have—previously lived in relative amity" (Appadurai 907). Conveying a wider sense of ecological and societal collapse through the contained and uncivil setting of Smalltown, Banerjee depicts the violence through mutilated appendages in an anthropophagic world where one's kin is one's meal.

In one scene, the Outsider rushes toward an abandoned primary school to find the village elders. Survivor 1 and Survivor 2 try to stop him and follow him to the school to ensure his safety. This is the first time we see the "monster" on screen (*Ghost Stories* 00:56:58)—a large, overgrown figuration resorting to cannibalism as a collective social process in Smalltown. The arrival of the monster on screen exudes a crippling sense of horror as it consumes Survivor 2, his daughter, in front of the Outsider and Survivor 1. With bestial and carnivorous affinities at display, the "monster" reminds the viewer that inflicting violence is not just a condition of the "social other"—it can also permeate to the "social own," reframing a nightmarish topology.[16] Therefore, to think that constructivist approaches and binary oppositions prevalent in the anthropocentric discourse trigger violence toward multiple "Other(s)" demonstrates an accurate but incomplete assessment of "violence" in posthuman discourse.

Violence is implicit in both inter- and intraspecies relationality. It goes beyond the "Human" and reflects the agentiality of matter, monsters, and more-than-human assemblages alike, signaling that the stories that we imagine can also become our imagined-collective reality while compromising the sustainability of our future(s). At this juncture, we become food for the monster and lose all agency in Banerjee's universe. The un(human)/monster acts as an apparatus of imagination and a cartographic function to explore the limits of being Human and the bond that we share with the natural and material world. In doing so, it dissolves the binaries of designated othernesses and others—the "naturalised other" (animals, biodiversity, planet), "racialised other," "sexualised other (women)," gendered other (LGBTQ+), and material other—who band together to delegitimize a dialectical scheme of being to produce a more-than-human discourse and storied ethics of care (Braidotti, *The Posthuman* 27). Therefore, there is a need to constantly speculate about and employ "figurative thinking" (Braidotti, *Feminism* 6).

A similar "storied" interlinkage is present in the writings of Barad and Jane Bennett, underlining the importance of "matter," defined as "lively, agentive and generative" (Oppermann 411). As a living, walking, devouring motif of ecophobia, Banerjee's "Monster" also acts as a situated articulation of "out-of-box thinking,"[17] which can be interpreted as "lively, agentive and generative" in an apocalyptic setup (Oppermann 414). Rob Nixon's suggestion of "rethink[ing]-imaginatively" is also relevant here (2). The insistence that "we can act" registers the urgency of times that we are living in and marks the presence of certain "unseen"[18] bodies at work (15). In addressing "violence," Nixon explicates,

> In a world permeated by insidious, yet unseen or imperceptible violence, imaginative writing can help make the unapparent appear, making it accessible and tangible by humanizing drawn-out threats inaccessible to the immediate senses. Writing can challenge perceptual habits and bring into imaginative focus apprehensions that elude sensory corroboration. The narrative imaginings of writer-activists may thus offer us a different kind of witnessing: of sights unseen. To allay states of apprehension—trepidations, forebodings, shadows cast by the invisible—entails facing the challenge, at once imaginative and scientific, of giving the unapparent a materiality upon which we can act. (15)

Consequently, we are constantly imagining new ways of situating the un(human) to expand the posthuman imaginary while foregrounding an intertwined existence of the human subject that evokes elation and apprehension synchronously, because "one [cannot just] continue to live but be nothing" (Muntean 84).

More specifically, from a posthumanist perspective, Banerjee's "Monster" raises a few fundamental questions: what happens when the world (as we know it) falls apart? Do we remain Human? Effectuating a grotesque imagery aligning with an apocalyptic framework, the "monster" forces Survivor 1 to eat Survivor 2's desecrated, blood-soaked remains and become "it"—a monster—to ensure his survival as "they [monsters] don't eat those who eat human flesh" (*Ghost Stories* 00:50:01). Informed by the claustrophobia induced by a charred landscape and the apocalyptic intentionality of dying or transmutating into a monster, we witness Survivor 1 eating Survivor 2's mutilated hand. The very next moment, he loses his sight, which confirms his altered existence as he enters the metamorphic zone of the un(human)/monster. Paul Sheehan sketches this metamorphosis as our ongoing crises as we are unsure of the beings that we are becoming—a fundamental lapse in the process of identifying with—and identifying as—potential signifiers of hybrid formations. Further, he elaborates,

> there is monstrosity all around us, which we refuse to see; but only that refusal protects us from our own worst tendencies. This is critical because the manufacture of monsters in recent years has been nothing short of prodigious. Just as the mythic body harks back to a fantastic past, and the techno-body upholds the promise of a radiant future, an obsession with monsters of every stripe—political, social and biological—defines our present-day condition. (Sheehan 258)

Therefore, by constructing the un(human)/monster, we dismember "its Humanist antipode Man" while deconstructing the structural hierarchization historically administered by the anthropocentric subject, pejoratively labeled as "unsanctioned," "monstrous couplings" by the Anthropos (Mazurov 263–64).

The Un(human)/Monster That We Require: Reading Hope

In considering the genealogy of Humanist universalism, the monster in its embodied configuration remains a marginal figure. Thus, by default, the monster becomes an icon, a cultural repository, and an anamorphic progeny in many posthumanist readings. In considering these conceptual encodings, Banerjee's vision delineates a dystopic re-imagination of "what we are as a collective whole," and constructs a relational configuration of the human and the monster through physical and physiological alteration, producing a liminal aesthetics of "what are you afraid of," and "what are you afraid for" (Banerjee 00:06:20). The idea of unsustainability and confronting harsh realities is implicit in Banerjee's apocalyptic imagination. Building on Mazurov's analysis of "Monster/The Unhuman," and the socio-cultural mapping of counternarrative produced by the affective alliance of seemingly unnatural trans-speciesism, we can confer that we need more un(humans)/monsters in our lives to unburden us from canonized models of anthropocentric bondage. More specifically, such hybrids project "ontological impropriety" (Braidotti, *Feminism* 218).

Furthermore, posthumanist readings engage in a creative exploration of post-anthropocentric figurations such as dragons, vampires, witches, insects, plants, viruses, cyborgs, and animals to generate alternate modes of articulation through speculation in a fictional world.[19] This counter-consciousness[20] is explored by Mazurov to decipher the intricate genealogy of the term "monster." He elaborates,

> The monster or the unhuman is not a negation but a non-relation, which is not to say that it is immune to situational interaction or relationality, but instead to highlight that the monstrous is a rejection of the stifling non/human binary entrapment. (Mazurov 261–62)

In this context, Banerjee's rendition of the cannibal and the cannibalized identifies a horrifying inevitability in the last shot of the film. The Outsider falls into a pit[21] while the monster jumps on top of him as he begs for his life. In the next frame, he is all alone as the un(human)/monster and all life forms disappear. The shock of confronting his impending death factualizes the ecophobic condition as the Outsider loses all semblance of the real. The Outsider reaches a tragic state of frenzy where we recognize human perishability and the transitory nature of life itself. With a dreadful deliriousness, the Outsider howls, "They ate the children; they ate the children!" (*Ghost Stories* 00:40:53). Through losing cognition and all sense of the real and the imagined, the Outsider has finally entered the un(human)/monster's mindscape. The apocalyptic subtext is further amplified as a few people abruptly show up from Bigtown to rescue the Outsider. The way they band together starkly resemble Alpha and his followers. They detest Smalltown. They tell the Outsider that there are no "humans" in Smalltown; it is a deserted place. The "Alpha" of this group acknowledges the Outsider's delirious state and says, "Don't worry. You're with us now. You are safe" (*Ghost Stories* 00:39:17).

Banerjee's "Monster" exemplifies the lost sense of belonging and our disjointed efforts to produce narratives of collective harmony and imaginings. Such on-screen depictions of life-threatening endemics reaffirm human vulnerability and the affect/balance required by our multi-species ecosystem. They reassert the necessity of forging inter-relational belongings by thinking beyond being-just-human. As certain more-than-human agents reaffirm "our" plural existence, we need to rethink our purpose through action, and "critical and creative" thought in this world (Braidotti, *Feminism* 150). We are transient. Therefore, as a response to the multiple ways of being interconnected with life and life forms beyond our reach, and the close relation to impending death and decay, especially post COVID-crises, now more than ever, we require speculative yet spectacular re-imaginations and narratives of doom to contemplate our allegiance with this planet because, "These are the worst of times, these are the best of times" (Braidotti, *Feminism* 242). In this regard, Dibakar Banerjee's "Monster" serves as a powerful testament to our collective articulations of becoming more-than-just human: a cautionary tale about our contemporary becoming(s).

Notes

1 In the compiled glossary of posthuman terms and themes edited by Rosi Braidotti and Maria Hlavajova (2018), Nikita Mazurov uses the terms "monster" and "unhuman" interchangeably and establishes them as relational entities. In borrowing Mazurov's terminology, this chapter aims to situate the "un(human)" as a self-reflexive metaphor for the "Human"—a subject under constant investigation in the posthumanist discourse.

2 Liminality or "in-betweenness" refers to the state of transition without initiation or completion—a place where transition is not informed by complete transformation. It highlights the concept of "becoming," rather than "being," in posthumanist discourse (Braidotti, *The Posthuman* 49). In the chapter, the term refers to our posthuman state of flux and dislocation from the Humanist ideal of knowledge production and a step toward post-anthropocentrism and inclusivity on a planetary level.

3 Braidotti, in her foundational work, *The Posthuman* (2013), refers to "the posthuman becoming" as "animal," "earth," and "machine" characterized by unprecedented variations of multiple, affective belongings (55–104). Since the "monster" in Banerjee's universe is produced through transformation and transmutability, the director admonishes the viewer to notice the intricate underpinnings of "becoming" a "monster" and a human rather than being one or both in an apocalyptic scenario.

4 See, for example, Hassan; Mazarov on the "monster" as an ever-evolving paradigm.

5 The term "contained" alludes to multiple meanings, including the stringent lockdowns imposed by many nations to mitigate the harrowing effects of COVID-19 pandemic, the idea of planetary singularity, where Earth is the only habitable planet, as well as the ideological "containment" of anthropocentric exceptionalism.

6 The term "agent" refers to the catalyst responsible for change. It can be human, nonhuman, more-than-human, fantastical, or material. The "social definition of 'actor'" by Latour is also important here to understand the pliability of the terms "agent" and "actor," which are used interchangeably in the chapter (374).

7 Manuela Rossini, in the essay "Bodies," from *Literature and the Posthuman* (2017), refers to the idea of forging a "coessentiality" where "being-with" or coexistence takes precedence in posthumanist discourse (154–57). She also refers to Jean Luc-Nancy's "singular plurality" as coexistence, which is the only essence that is essential for survival (157). Nancy reiterates that, "co-essence, or being-with (being-with-many), designates the essence of co-" (*Being Singular Plural* 30; Rossini 157). Here, "co-essence" raises a fundamental question: if the co-"other" is the monster, then who are we?

8 The term is a spatiotemporal metaphor used to denote the urgency of our contemporary times and the need to act now, at this very moment.

9 In a promotional interview, Banerjee uses the phrase, "imagine a society where I can… make you into a nice ham sandwich" to describe the intentionality behind the film (*The Quint* 00:01:07).

10 By re-reconfiguring the process of "becoming" as an infinite flow of multiple "deterritorialized-becoming," and embracing our subjective "discontinuity," and "displacement," we enter a state of "deterritorialized flux" (Deleuze and Guattari 13). Becoming-human, here, then, refers to a revived understanding of a complex interconnected state that is the present moment, especially post-COVID-19 new normalcy, situating humane workings on a spectrum of vulnerabilities and an ethics of deep care extended toward more-than-human narratives.

11 The term "anxieties," like the term "monster," acts as a cultural repository. Our collective "anxieties" inform our daily remedial actions toward the planet and confirm our collective incapacity to produce effective change.

12 We can extend this metaphor to the actuality of COVID-19 crises where the fear of the contagion, the passivity toward human touch, and uninhibited mobility shaped a new reality or the new normal.

13 The experience and notion of a "safe" home or space has been debated across disciplines. Post-COVID-19 pandemic, sanctioned lockdowns, and factors contributing to climate change, the idea of "home" has undergone significant alterations, disruption, and destabilizations. Conrad Scott introduces the prefix "eco-" to "critical dystopia" to elucidate the significance of a "dwelling," and a "sense of place" that can be "tangibly extrapolated forward from a present moment in the real world" (213). The spatial erasure, geographical migration, and locus ambivalence during unprecedented events often result in an uncaring relationship with our environments, and "what might be called home" (Scott 213).

14 The phrase "fantastic fabulation" is derived from Donna Haraway's concept of "speculative fabulation." It draws inspiration from multiple discourses such as science fiction, gothic, horror, and fantasy literature. It refers to the sense of collective responsibility endowed upon us to create a sustainable future by maintaining an ecological balance. "Fantastic fabulation" allows posthuman icons such as monsters, ghosts, witches, animals, and uncanny motifs to act as navigational tools to decipher and critically disband binary oppositions, exclusionary politics, and anthropocentrism (Braidotti, *Posthuman Feminism* 213).

15 The act of "being" from an eco-philosophical axis entails an egotistical presumption of a unified being and sovereignty over all others. However, "becoming" allows for a much more inclusive, interconnected web of relations and socio-cultural, material exchange in asserting collective interactions, vulnerabilities, dualities, and challenges issued by the "Anthropos."

16 The phrase "social other" bears the undercurrents of anonymity or unfamiliarity. Here it is used as an antonym of "socially proximate persons," whereas the "social own" refers to kin.

17 Out of box thinking according to Oppermann refers to moments in crises in an apocalyptic scenario where we are confronted with our worst anthropocentric anxieties or reminded of the false sense of exceptionality projected by the "Humanistic Human."

18 The "unseen" refers to a broad range of "agents" such as bacteria, viruses, toxicity, decomposition, pollution, etc.

19 *Fantastic Beasts and Where to Find Them*, directed by David Yates (2016), *Raya and the Last Dragon*, directed by Carlos Lopez Estrada and Don Hall (2021), *Sherni*, directed by Amit V. Masurkar (2021), *Stree*, directed by Amar Kaushik (2018), *Bhediya*, directed by Amar Kaushik (2022), *Morbius*, directed by Daniel Espinosa (2022), *Avatar: The Way of Water*, directed by James Cameron (2022), *Wednesday*, directed by Tim

Burton (2022), and *Tooth Pari: When Love Bites*, directed by Pratim Dasgupta (2023), are some of the recent stories that explore more-than-human figurations along with relevant critique of our present injustices.

20 Mazurov places monster/un(human) as a counter measure to the "Human" with its Humanist baggage.

21 The fall into the pit signifies the breakdown of "human supremacy and exceptionalism," when juxtaposed with the un(human), and thus acts as a posthuman metaphor. The Outsider's actual fall into the literal pit refers to a loss of human agency and a profound disruption of the not-so-ethical foundations of the age-old anthropocentrism. We see a reversal of the chain of hierarchy which places humans at the center of all life forms. Alpha looks down upon the Outsider as he finally becomes food to satisfy its hunger.

Works Cited

Appadurai, Arjun. "Dead Certainty: Ethnic Violence in the Era of Globalization." *Development and Change*, vol. 29, no. 4, 1998, pp. 905–25.

Barad, Karen. *Meeting the Universe Halfway: Quantum Physics and the Entanglement of Matter and Meaning*. Durham, NC: Duke UP, 2007.

Braidotti, Rosi. *Metamorphoses: Towards a Materialist Theory of Becoming*. Cambridge: Polity P, 2002.

———. *Posthuman Feminism*. Cambridge: Polity P, 2022.

———. *Posthuman Knowledge*. Cambridge: Polity P, 2019.

———. *The Posthuman*. Cambridge: Polity P, 2013.

Braidotti, Rosi, and Maria Hlavajova, editors. *Posthuman Glossary*. Bloomsbury Academic, 2018.

Cohen, Jeffrey Jerome, editor. *Monster Theory: Reading Culture*. U of Minnesota P, 1996.

Deleuze, Gilles, and Felix Guattari. "Kafka: Toward a Minor Literature: The Components of Expression." *Theory and History of Literature*, vol. 30, U of Minnesota P, 1986.

Estok, Simon C. "Cannibalism, Ecocriticism, and Portraying the Journey." *CLCWeb: Comparative Literature and Culture*, vol. 14, no. 5, 2012. https://doi.org/10.7771/1481-4374.2138

———. *The Ecophobia Hypothesis*. Routledge, 2018.

Ferrando, Francesca. "Being Posthuman Now." *After Human? Making Sense between Art, Science, and Religion in the 21st Century*, Interactive Lecture. U of Linz, 17 January 2023, https://www.theposthuman.org/austria-2022.html.

Haraway, Donna J. *The Companion Species Manifesto: Dogs, People, and Significant Otherness*. Prickly Paradigm P, 2003.

———. *When Species Meet*. U of Minnesota P, 2013.

Hassan, Ihab. "Prometheus as Performer: Toward a Posthumanist Culture?" *The Georgia Review*, vol. 31, no. 4, 1977, pp. 830–50. *JSTORE*, http://www.jstor.org/stable/41397536.

Heise-von der Lippe, Anya, editor. *Posthuman Gothic*. U of Wales P, 2017.

Iovino, Serenella, and Serpil Oppermann, editors. *Material Ecocriticism*. Indiana UP, 2014.

Jean-Luc, Nancy. *Being Singular Plural*. Stanford UP, 2000.

Kermode, Frank. *The Sense of an Ending: Studies in the Theory of Fiction*. Oxford UP, 2000.

Latour, Bruno. "On Actor-network Theory: A Few Clarifications." *Soziale Welt*, vol. 47, no. 4, 1996, pp. 369–81. *JSTOR*, http://www.jstor.org/stable/40878163.

Mazurov, Nikita. "Monster/ The Unhuman." *Posthuman Glossary*, edited by Rosi Braidotti and Maria Hlavajova, Bloomsbury Academic, 2018, pp. 261–64.

"Monster," *Ghost Stories*. Directed by Dibakar Banerjee, performances by Sukant Goel, and Gulshan Devaiah, *Netflix*, 1 January 2020, https://www.netflix.com/in/title/81088083.

Muntean, Nick. "Nuclear Death and Radical Hope in *Dawn of the Dead* and *On the Beach.*" *Better off Dead: The Evolution of the Zombie as Post-human*, edited by Deborah Christie et al., Fordham UP, 2011, pp. 81–97.Nayar, Pramod K. "Appupen's Posthuman Gothic: The Snake and the Lotus." *South Asian Review*, vol. 39, no. 1, 2018, pp. 70–85. https://doi.org/10.1080/02759527.2018.1515802

Nixon, Rob. *Slow Violence and the Environmentalism of the Poor*. Harvard UP, 2011.

Oppermann, Serpil. "Storied Matter." *Posthuman Glossary*, edited by Rosi Braidotti and Maria Hlavajova, Bloomsbury Academic, 2018, pp. 411–14.

Oziewicz, Marek. "Introduction: The Choice We Have in the Stories We Tell." *Fantasy and Myth in the Anthropocene: Imagining Futures and Dreaming Hope in Literature and Media*, edited by Marek Oziewicz et al., Bloomsbury Academic, 2022, pp. 1–11.

Parui, Avishek, and Merin Simi Raj. "The COVID-19 Crisis Chronotope: The Pandemic as Matter, Metaphor and Memory." *Memory Studies*, vol. 14, no. 6, 2021, pp. 1431–44.

Rossini, Manuela. "Bodies." *Literature and the Posthuman*, edited by Bruce Clarke and Manuela Rossini, Cambridge UP, 2017.

Scott, Conrad, "Post-Anthropocenic Undying Futures: The Ecocritical Dystopian Posthuman in Lai's *The Tiger Flu* and Bacigalupi's "The People of Sand and Slag."" *The Anthropocene and the Undead: Cultural Anxieties in the Contemporary Popular Imagination*, edited by Simon Bacon, Lexington Books, 2022, pp. 211–26.

Sheehan, Paul. "Posthuman Bodies." *The Cambridge Companion to the Body in Literature*, edited by David Hillman et al., Cambridge UP, 2015.

The Quint. "Dibakar Banerjee and Gulshan Devaiah on Netflix's *Ghost Stories*." *YouTube*, uploaded by *The Quint Channel*, 26 December 2019, https://www.youtube.com/watch?v=708cBcISkdw.

———. "Karan Johar, Anurag Kashyap, Zoya Akhtar and Dibakar Banerjee Talk About *Ghost Stories*." *YouTube*, uploaded by *The Quint Channel*, January 2020, https://www.youtube.com/watch?v=HURdRedXfTg. Accessed 12 December 2022.

3 A Scourge Even Worse Than Disease

Richard Matheson's *I Am Legend* as Pandemic Political Allegory

Timothy S. Murphy

Il y a des héros en mal comme en bien.
[Evil, like good, has its heroes.]
La Rochefoucauld, *Maximes* #185

Richard Matheson's novel *I Am Legend*, which focuses on the lone human survivor of a bacterial plague that transforms both living and dead people into vampires, has remained continuously in print since its original publication in 1954. It has also inspired three feature film adaptations—*The Last Man on Earth* with Vincent Price (1964), *The Omega Man* with Charlton Heston (1971), and *I Am Legend* with Will Smith (2007)—as well as George Romero's *Night of the Living Dead* (1968), and through the latter, it has helped define the enormous sector of the culture industry dedicated to zombie narratives over the past 50 years.[1] Even so, readers today, having experienced the COVID-19 pandemic, may be better able to empathize with protagonist Robert Neville's physical, affective, and social isolation inside his fortified Southern California home than many previous audiences. Matheson describes Neville's psychological condition in intensely claustrophobic, indeed paranoic terms that resonate not only with the contemporary experiences of lockdown, self-quarantine, and conflicting or unreliable information regarding infection but also with the increasingly crude polarization of global political discourse, particularly the erosion of representative democracy, that accelerated during the pandemic. This erosion has reached one of its nadirs (so far, anyway) with former US President Donald Trump's refusal to accept his re-election loss, in part because of changes in voting methods spurred by the pandemic. Trump's ousting from office has by no means reversed, or even slowed, that erosion, as events taking place in supposedly democratic regimes all over the world continue to show. Matheson's novel, published at the outset of the Cold War and on the cusp of the modern Civil Rights movement's emergence, invites us to read it as an allegory of such political transformations, as this chapter will argue.[2]

The novel in question is divided into four parts of unequal length spanning the three years from January 1976 (perhaps not coincidentally the start of the US Bicentennial celebration, still more than 20 years in the future at the time of

DOI: 10.4324/9781003345770-5

the novel's publication) to January 1979, in the medium-term aftermath of an unstoppable plague or pandemic that struck the United States—and presumably the world—in 1975. The plague's origin is never fully explained, in that the bacterial pathogen causing the rapidly spreading mass infections is not scientifically identified until long after civilization's collapse, when the increasingly desperate Neville trains himself to use a microscope and isolates it, dubbing it "*vampiris*" (Matheson 75).[3] In keeping with the novel's early Cold War context, both radioactive fallout and germ warfare are suggested as possible origins for the pandemic (44–46); this uncertainty uncannily foreshadows the ongoing debate today over the origin of the COVID-19 virus. The infection itself represents quite a unique intervention into the genre of vampire narratives in that it is bivalent: in some cases, it kills its human hosts and restores them to a kind of life as undead vampires, while in other cases, it simply modifies the metabolism of living humans so that they must consume blood to live, although they do not actually die of the infection. This second valence, which may strike contemporary readers as analogous with "long COVID," is crucial to the story's outcome. Part One of the novel introduces Robert Neville, his besieged home in the south-central Los Angeles suburb of Gardena, and his undead vampiric adversaries, led by his former neighbor and co-commuter Ben Cortman. Part Two, set two months later, narrates Neville's slow but ultimately successful efforts to understand the plague's vampiric symptoms scientifically, as well as his desperate and ultimately unsuccessful attempt to bring a terrified, injured dog into his home for companionship. Part Three jumps ahead to June 1978, when Neville, now apparently resigned to his isolation, meets an apparently uninfected woman named Ruth and forcibly drags her back to his home for interrogation and testing. This part concludes when she knocks him unconscious and escapes, leaving behind a note informing him that a group of infected living humans has survived and is organizing itself to take the city back both from the undead vampires and from Neville himself, whom they view as a threat comparable to the undead. Part Four presents Neville's final day, six months later, when the organized survivors come for him.

The present analysis focuses mainly on the first and last parts, the novel's formal and thematic frame, with a few detours into Parts Two and Three, especially the scenes that lay out the terms and patterns for a political allegory. By that, I mean a narrative in which individual characters and situations can be mapped onto broader political categories and practices that they explicitly or implicitly embody or represent, without thereby ceasing to show psychological depth or complexity. An explicit instance of political allegory, perhaps the most important one for my argument here, appears in Chapter 3 of *I Am Legend*, as the undead Cortman calls Neville out by name and unnamed vampire women appeal to his barely controlled sexual desire by flaunting themselves nude before his door. Against this backdrop, Neville debates with himself why he doesn't simply stop resisting and join them, which is the only "sure way to be free of them" (18). Although he can imagine that other uninfected survivors like himself exist, he has no way to locate, contact, or join them, so the vampires paradoxically appear to be his only possibility for social contact and interaction. The debate concludes with a self-consciously formal—and heavily ironized—forensic defense of the "thesis" that "Vampires are prejudiced against" (20). Neville begins by identifying the "keynote of minority prejudice…: They are loathed because they are feared" (20). He then launches into a list of comparisons intended to refute the claim that vampires' "needs [are] any more shocking than the needs of other animals and men," a list that includes such recognizable—and socially tolerated—types as "the parent who drained the spirit from his child" or who "gave to society a neurotic child who became a politician," "the manufacturer who set up belated foundations with the money he made by handing bombs and guns to suicidal nationalists" (a veiled reference to chemist Alfred Nobel's establishment of the Nobel Prizes in 1895), "the publisher who filled ubiquitous racks with lust and death wishes" (a category that might well include this novel itself), and others (20–21).

Neville's summation for the defense concludes with a peroration that explicitly evokes the conflicts surrounding the US Supreme Court's *Brown v. Board of Education* decision striking down racial segregation in education, which had been argued before the Court in spring and fall 1953 but was announced a scant three months before *I Am Legend* was first published in August 1954:

> Why, then, this unkind prejudice, this thoughtless bias? Why cannot the vampire live where he chooses? Why must he seek out hiding places where none can find him out? Why do you wish him destroyed? Ah, see, you have turned the poor guileless innocent into a haunted animal. He has no means of support, no measures for proper education, he has not the voting franchise. No wonder he is compelled to seek out a predatory nocturnal existence. (21)

Using the classic method of ideology critique best exemplified by Frederick Douglass' *Narrative of the Life* (1845), Neville reveals the inversion of cause and effect that serves as the basis of contemporary prejudice just as it had served as the basis for the defense of chattel slavery. The legal and social restrictions imposed upon the minority or outsider in supposed defense of the majority are not responses to pre-existing characteristics or threats, but in fact the causes of those very threats. The references to the minority person's inability to "live where he chooses" and to the foreclosure of the "voting franchise," legal issues that were not addressed in *Brown v. Board of Education*, look forward to further civil rights struggles a decade later, culminating in the Civil Rights Acts of 1964 and 1968 and the Voting Rights Act of 1965. This apparent expression of sympathy and appeal for tolerance, couched in the language of political progressivism, is immediately undercut, however, when Neville thinks, "Sure, sure…but would you let your sister marry one?" and responds to himself, "You got me there, buddy, you got me there" (21).[4]

Neville's question, of course, has been a standard challenge raised by segregationists and white supremacists against proponents of racial equality and integration for well over a century, and it is often asked in that same quasi-rhetorical fashion, indicating a negative answer is expected. One of the earliest and most startlingly explicit examples of such an exchange in print occurs in Chapter 10 of James Weldon Johnson's classic 1912 novel of passing, *The Autobiography of an*

Ex-Colored Man, and although no evidence exists that Johnson directly influenced Matheson, the parallelism between their respective scenes is nonetheless striking. Johnson's light-skinned narrator, traveling by train through the American South to collect orally circulating African American folk music for use in the written musical compositions he is planning, overhears a debate between an ex-Union soldier and a Texas cotton planter over the "race question" (Johnson 82) in the United States at the end of the nineteenth century. After the old soldier demolishes the Texan's ignorant pretension that "[t]he Anglo-Saxon race has always been and always will be the masters of the world, and the niggers in the South ain't going to change all the records of history" (Johnson 84–85), the planter makes his final plea: "You want us to treat niggers as equals. Do you want to see 'em sitting around in our parlors? Do you want to see a mulatto South? To bring it right home to you, would you let your daughter marry a nigger?" At this the old soldier demurs, just as Neville does, and in so doing grants his opponent's larger point: "No, I wouldn't consent to my daughter's marrying a nigger, but that doesn't prevent my treating a black man fairly" (Johnson 85).

Wherever Matheson found inspiration for it, the self-debate scene constitutes Neville's most thorough critical examination of his own ambivalent attitudes toward the vampires prior to the conclusion of the novel, although it concludes by re-affirming his own fear and loathing in terms that would have carried obvious political resonance in 1954. The rest of Parts One and Two narrates his intensifying solitude, the contraction of his intersubjective horizon, and the concomitant gradual coarsening and hardening of his ethical judgment, which inversely and ironically parallel the growth and refinement of his medical understanding. Initially, his self-imposed mission to eliminate all vampires, both living and undead, from his surroundings causes him some psychological trauma: "Usually he felt a twinge when he realized that, but for some affliction he didn't understand, these people were the same as he" (28). After a few more weeks of physical and psychological isolation, though, he accustoms himself to the task, first by consciously suppressing his former doubts, and later by fully repressing them: he began "to suspect his mind of harboring an alien. Once he might have termed it conscience. Now it was merely an annoyance. Morality, after all, had fallen with society. He was his own ethic" (50). This observation seems a remarkable premonition of how the COVID-19 pandemic would further polarize and balkanize already fractured ethical and political discourses, both in the United States and elsewhere. In the novel, the disintegration of social solidarity and its associated morality seems to have begun with Neville's decision, at the height of the pandemic months earlier, to break the emergency law requiring the immediate cremation of all infected bodies by secretly burying his wife Virginia following her death from the disease. This act is witnessed and denounced by an unnamed neighbor:

> "I'm not going to the fire, I said!" Neville blurted out, and jabbed in the starter button.
> "But your wife," the man said. "You have your…"
> …

> "I'm not *going* there!" Neville shouted without looking at the man.
> "But it's the *law!*" the man shouted back, suddenly furious. (63)

Despite Neville's effort to inter her according to pre-pandemic traditions, Virginia soon returns home as an undead vampire whom he must dispatch, an event that inflicts permanent psychological trauma upon him (65–66; see also Christie 72–74). However, the experience doesn't prevent Neville from continuing to set his own assumptions and preferences above empirical evidence of how the world has changed. What begins as a personal moral choice or private ethical stance in conflict with the juridical law (and with the medical principles on which that law is based) soon develops into a new law unto itself, of which Neville is legislator, judge, jury, and executioner. In the case of the COVID-19 pandemic, early communications errors on the part of medical authorities such as the Centers for Disease Control and Prevention (CDC) led first to confusion and soon thereafter to doubt and hardening resistance on the part of a rapidly coalescing group of "denialists" and "anti-vaxxers" who disputed the disease's fatality rate and the vaccine's efficacy. On the basis of those disputes, they refused to perform many of the measures intended to slow the spread of the infection. Such denialism empowered them to reject all appeals to the public good in favor of claims to personal freedom utterly without social responsibility, just as Neville's did.

Part Two concludes with Neville's recollection of his unwilling participation in a raucous tent revival shortly after re-burying his wife. To the hysterical crowd the preacher shouts, "I ask you—do you want to be turned into godless, night-cursed husks, into creatures of eternal damnation?" They reply, "terror-stricken, 'No, *save* us!'" But the preacher only repeats, "We are damned, we are damned, *we are damned!*" (103). This memory not only helps Neville explain to himself the vampires' avoidance of religious symbols such as crosses or the Torah as a purely psychological response to their own self-loathing as "creatures of eternal damnation," but it also helps him find a degree of acceptance of and even pleasure in his own lonely situation and retributive project:

> I don't have to escape from anything. Not now.
> For the first time since the dog had died he smiled and felt within himself a quiet, well-modulated satisfaction. There were still many things to learn, but not so many as before. Strangely, life was becoming almost bearable. I don the robe of hermit without a cry, he thought. (107)

The juxtaposition of the vampires' fear of damnation with Neville's newfound, pseudo-monastic commitment to their destruction has the added consequence of heightening the irony underlying the ethical reversal that concludes the novel.

At the beginning of Part Three, an ascetic yet physically "bigger, more relaxed Neville" encounters an apparently uninfected woman, Ruth, who emerges from the same field "where he had buried [his wife Virginia], where she had unburied herself" (109). Thus, her appearance initially hangs suspended between two figural alternatives that force a choice or decision upon Neville: is she a replacement for the wife he has lost, or is she another vampire coming for him? In other words, will this meeting result in the fulfillment of Neville's fantasy of restoring the lost world order, with Neville and Ruth as new Adam and Eve (114–15), or will it end in yet another killing as blood tribute to that lost order? The ultimate answer to this fundamentally allegorical question will be both, and neither. When Ruth flees from him, he chases and catches her, asking what she's afraid of. "He didn't realize that his voice was devoid of warmth, that it was the harsh, sterile voice of a man who had lost all touch with humanity" (113). He forcibly carries her to his fortress-home in order to test her blood for infection, and throughout the process alternates between hope or desire for companionship and reflexive suspicion of her status. Only later does he begin to suspect the extent to which this procedure manifests his own increasing dehumanization, realizing that, "Like Gulliver returning from the logical horses, I find the human smell offensive" (126). Ruth's expression of horror at his account of destroying both living and undead vampires perplexes him: "Horrible? Wasn't that odd? He hadn't thought that for years. For him, the word 'horror' had become obsolete. A surfeiting of terror soon made terror a cliché. To Robert Neville the situation merely existed as natural fact. It had no adjectives" (134). He no longer feels any "twinge" when he kills. When Ruth asks him how he knows that the living vampires aren't going to "stay alive," he answers with absolute certainty that, "No matter how long their systems fight it, in the end the germ will win.... If I didn't kill them, sooner or later they'd die and come after me. I have no choice; no choice at all" (135).

This information appears to answer the unasked question lying behind Ruth's spoken one, since she escapes from Neville the next morning, following a brief moment of emotional intimacy, by striking him down with the very mallet he uses to stake vampires (12, 142). She leaves behind a note explaining that she's actually a spy sent to gather information about him. She had volunteered for this job, in part, because Neville had killed her husband, who, like her, was not an undead vampire but a living human infected with the *vampiris* bacterium. She is part of an organized group of infected living humans that has resolved

> to stay alive. We've found a way to do that and we're going to set up society again slowly but surely. We're going to do away with all those wretched creatures whom death has cheated. And, even though I pray otherwise, we may decide to kill you and those like you. (143)

This revelation stuns Neville, as he had long ago concluded about both kinds of vampire that "There was no union among them. Their need was their only motivation" (11). His use of the loaded term "union" here implicitly marks his thinking as political as well as social. Ruth urges him to flee the city before they decide to eliminate him, and she leaves him one of the pills that "saved us from dying, that is helping us to set up society again" (144). When he examines the pill through his microscope, it reveals to him both his intellectual failure to foresee that "Bacteria can mutate" (146), thereby possibly becoming susceptible to the vaccines he had failed to develop in Part Two, and, by implication, his moral or ethical failure to differentiate between the living vampires and the undead. Contrary to his own stated conclusion, he did have a choice as to whether to kill indiscriminately or not. As Deborah Christie asserts, "[H]e has failed to recognize his own too-narrowly-defined classification for humanity, and in his error he has been the agent of humanity's destruction" (Christie 76).

When the new humans or posthumans, that is, the living vampires who control their infections with the pills, come for him months later, Neville is appalled by the barbarity of the mass extermination techniques they deploy against the undead vampires, chiefly automatic rifles and long pikes, though these seem like straightforward extensions of his own much slower serial methods using pistols and wooden stakes. After all, collective violence is one of the earliest and most predictable consequences of collective organization. This new and unexpected trauma abruptly forces him to acknowledge his own unconscious identification with their undead victims: "He didn't want them to get Cortman, he realized, didn't want them to destroy Cortman like that. With a sense of inward shock he could not analyze in the rush of the moment, he realized that he felt more deeply toward the vampires than he did toward their executioners" (147–48), perhaps because the undead don't pose the ethical problem for him that the living do, or rather should have done if he hadn't successfully disavowed and repressed it. Although he originally intends to submit peacefully to his captors, at the last moment he panics and decides instead to resist, takes a bullet to the chest, and blacks out, after which they drag him out "Into the night. Into the world that was theirs and no longer his" (152). When he slowly awakens in what is evidently a cell, his pain reinforces that identification with the undead, but also implicitly echoes the suffering of the living he had killed: "This is what they must have felt when the pikes went into them, he thought. This cutting, biting agony, the escape of life's blood" (154).

In the final scene (which none of the film versions has staged without major changes, though the 1964 version comes closest to accuracy), Ruth visits Neville to explain his situation and the fate that awaits him. She reveals that, in addition to working as a spy, she also plays an overtly political role as a leader of the infected living humans, who have decided that he must be executed as punishment for what they see as his crimes. She and her comrades are, she says, "like a revolutionary group—repossessing society by violence" (155). To his accusation that they take too much joy in killing the undead, she replies, "Did you ever see *your* face…when you killed? …I saw it—remember? It was frightening. And you weren't even killing then, you were just chasing me" (156). His violence against both types of vampires is nothing more than a slow and piecemeal version of the very same violence Ruth's comrades inflict on a collective, almost industrial scale. Although the ethical equivalency that Matheson implies here between the large-scale, rapid, "revolutionary" or spectacular violence practiced by Ruth's comrades and the small-scale, slow-motion, comparatively intimate violence of Neville's reactionary one-man mission does not precisely anticipate the "slow violence" of environmental degradation, neo-colonialism, racialized policing, and related aspects of neoliberal development famously theorized by Rob Nixon (2011), it does highlight the continuity or complicity among forms or degrees of violence in a way that seems compatible with Nixon's analysis. Heretofore Neville has always considered himself "a man, not a destroyer" (128), but after she informs him that he's the last uninfected human—"When you're gone, there won't be anyone else like you within our particular society" (156)—Neville finally realizes that he cannot be a heroic defender of ethical right and civilization (and thus the inspiration for scores of later zombie hunters), nor is he true humankind's last noble martyr. Instead, he recognizes that he can only be the first new monster that will haunt the new, permanently infected posthuman species—*Homo vampiris*?—as it (re)builds civilization:

> And suddenly he thought, I'm the abnormal one now. Normalcy was a majority concept, the standard of many and not the standard of just one man.... To them he was some terrible scourge they had never seen, a scourge even worse than the disease they had come to live with. He was an invisible specter who had left for evidence of his existence the bloodless bodies of their loved ones. And he understood what they felt and did not hate them. (159)

He had hated them, loathed them, because he feared them, but here he recognizes that his fear, at least with regard to the living vampires, was misplaced, and they had more to fear from him than he did from them. To put it bluntly, "Neville has become the monster and the vampires have become representatives of the post-human" (Christie 76). In the eyes of these "new people of the earth," he was "anathema and black terror to be destroyed," the type specimen for *Homo sapiens* itself no longer viewed as the moral and intellectual crown of creation but rather as "A new terror born in death, a new superstition entering the unassailable fortress of forever" (173–74). He has exchanged symbolic roles and places with the now-ascendant *Homo vampiris*.

The novel's concluding sentence, uttered by Neville to himself as the poison Ruth smuggled to him takes effect, finally pronounces its evocative title, "I am legend." As the culmination of Neville's dying reflection, this phrase echoes and resonates with the great ironic inversions of conventional ethical values scattered across literary history, especially those that are presented as perverse expressions of solace. Among those, we must number not only La Rochefoucauld's sardonic maxim on the possibility of heroic evil, with which this chapter begins, but also Satan's apostrophe in book four of John Milton's *Paradise Lost*: "So farewell Hope, and with Hope farewell Fear,/Farewell Remorse: all Good to me is lost;/Evil be thou my Good" (Milton ll. 108–10), which Victor Frankenstein's creature later cites in its own final soliloquy (Shelley). Slightly closer to Matheson's American Cold War context is Huckleberry Finn's famous decision to "*go* to hell" rather than send his friend Jim back into slavery in Chapter 31 of Mark Twain's 1884 novel:

> It was awful thoughts and awful words, but they was said. And I let them stay said; and never thought no more about reforming. I shoved the whole thing out of my head, and said I would take up wickedness again, which was in my line, being brung up to it, and the other warn't. And for a starter I would go to work and steal Jim out of slavery again; and if I could think up anything worse, I would do that, too; because as long as I was in, and in for good, I might as well go the whole hog. (Twain)

In this passage, Huck does not reject or criticize his society's racist values in the name of some other, more humane or progressive values, but simply accepts that, since he himself is "wicked" from birth, his instinctive unwillingness to condemn Jim to further enslavement can only be the final proof of his own moral failure, which he actively embraces. Matheson's scene is suffused with a mode of irony that parallels but ultimately differs from Twain's, though on the whole they both allegorize the political conflict now being fought (as it has been fought at irregular intervals for centuries) over the legal and logical cornerstone of representative democracy, that is, the rights of a minority living under a majority's rule, including the right of that minority to become a majority. Twain's reader from the Gilded Age and after can see what Huck, trapped in the antebellum American South, cannot, that his instinctive opposition to slavery represents an ethical majority yet to come; Matheson's reader, on the other hand, is compelled to share in Neville's insight that he represents an ethical majority doomed to vanish.

Furthermore, Neville's finale also effectively allegorizes the psychological struggle of the former majority to retain power as it descends, unwillingly, into minority status.[5] Neville's inability—or, less charitably, his refusal—to distinguish between infected living persons and undead revenants blinds him to the emergence of the former's new social and political order and its claims to majority rights. His increasingly complacent and "satisfying" monastic isolation deadens his few remaining ethical or humane reflexes, which, in turn, authorizes his indiscriminate destruction of both infected types, in the indisputably absurd hope of restoring the pre-pandemic order. These characteristics also reveal him to be not unlike the deluded twenty-first-century "patriots" who disdain masks, vaccines, and social distancing while investing full faith and credit in an impossible project of historical reversion to a past that never really was. We might even label him—and maybe them—political serial killers, by which I mean an atavistic personification of blind and obsolete majoritarian barbarism in one of its most irrational, self-reinforcing, and destructive forms. Acts Neville presumes to be ethically good are worse than the plague's evils from his victims' viewpoint; he is not merely an anachronism or obstacle to their rebuilding efforts but a merciless counterforce, an agent of slow violence only because he lacks the means to speed it up. The final unfolding of the plague's long-term consequences forces Neville to recognize that he is no longer part of a majority, but rather a minority of one, and like so many other uncomprehending minorities, he clings to his inevitably waning power more furiously, violently, and blindly as it slips away. This isolation is finally brought home to Neville when Ruth tells him that he is "the last of the old race" (156). Unlike many majorities in the process of becoming minorities, however, he is granted a last-minute epiphany or self-revelation, a private and personal apocalypse within the global and public apocalypse—in the original meaning of that term: an unveiling or flowering of what has heretofore remained concealed[6]—that permits him to recognize and momentarily sympathize with the posthuman successors whom he has driven into the role of conquerors.

His self-defeating acts, if not his belated recognition of their real meaning, establish Neville himself as himself a kind of politically allegorical vampire or revenant—a metaphorically undead personification of self-justifying anti-democracy and minority rule, of a piece with the reactionary social and political positions and groups such as the "Great Replacement" theory and Q-Anon that found new opportunities to assert themselves during the COVID-19 pandemic. Those positions and groups have a long history that has allowed them to sink deep roots into almost every nation-state. In a US context, Neville's narrative role is symbolically compatible with the now-defunct Three-Fifths Compromise over slavery (annulled by the Fourteenth and Fifteenth Amendments) and the still functional US Senate and Electoral College (which both subordinate the representation of people or communities to the representation of state institutions) that were all written into the US Constitution in 1787 (*Constitution of the United States of America*, article 1, sections 2 and 3, and article 2, section 1) in order to ward off "mob rule." Moreover, his restatement of racist defenses of white supremacy, however ironically it might be intended, echoes the sectional minority veto that John C. Calhoun called the "concurrent majority" and hoped would permanently preserve Southern slavery from Northern abolitionism (Calhoun 35–40, 45–49). Neville's failures also resonate with the more recent series of late nineteenth-century US Supreme Court cases, culminating in *Plessy v. Ferguson* (1896), that authorized domestic racial segregation until they were overturned by *Brown v. Board of Education*, as well as the Immigration Act of 1924 that effectively defined and defended the US as a white nation. Beyond the United States, the constitutive pseudo-scientific discourses of Anglo-European imperialism (which Rudyard Kipling figured as the "white man's burden" in his 1899 poem with that title), along with their belated Asian imitators such as Japan, which were in the process of disintegrating even as Matheson was writing this novel, attempted to preserve a vanishing pseudo-majority hegemony in increasingly desperate and obviously self-contradictory ways that bear obvious resemblances to Neville's self-justifications.

But as we have seen, Neville's paranoia and anxiety in the face of otherness, as well as the dehumanization of both self and other that those feelings triggered, all of which were expressed through murderous violence, were in no way rendered obsolete by the military defeat of fascism, the dismantling of colonial empires, or the increasing acknowledgment of minority rights within nation-states and even by the United Nations after the Second World War. The reactionary roots remain even after the flowers have been pruned. Neville's fears foreshadow the global revival and proliferation of anti-democratic mass politics after the wave of colonial independence movements, the equally radical global interregnum of the 1960s, and the purported end of collective antagonisms following the collapse of the Soviet Union in 1991, an event prematurely labeled the "end of history" by neoconservative theorists such as Francis Fukuyama (1992). The specific contexts and forms of such contemporary anti-democratic movements are legion, but they fall into a few general categories that resonate to varying degrees with aspects of Neville's self-justifications and actions. Alongside the comparatively conventional non-democratic authoritarianisms of Vladimir Putin's Russia, Xi Jinping's China, and Kim Jong-un's North Korea, some of which deploy sham elections to legitimize single-party rule, we should also recognize the formally "democratic" but decidedly "illiberal" authoritarianisms of Jair Bolsonaro's Brazil, Rodrigo Duterte's Philippines, and Viktor Orbán's Hungary as well as the majoritarian ethnonationalisms that have intermittently prevailed in Cambodia, Bosnia, Croatia, Serbia, Sudan, and many other states (including, in my view, the slow-motion catastrophe that is the post-Brexit United Kingdom); the more systemically theocentric nationalisms of Myanmar, Iran, Taliban-controlled Afghanistan, and the states of the Arab Peninsula; and the hybrid ethno-theo-nationalisms of Pakistan, India, Türkiye, and Israel, all of which currently serve as structural models for the increasingly overt white Christian nationalist movement in the United States and parts of Europe. All of these otherwise different cases share one prominent feature: each one seeks to use the ideology and trappings of pluralist democracy, namely the claim to majority legitimacy, to preserve the political power of shrinking numbers of traditionally defined citizen-subjects and prevent new types of citizen-subjects—women, LGBTQIA subjects, people of color, disabled people, religious minorities, non-conformists, and non-believers—from acceding to the political participation their increasing numbers would otherwise require. While the COVID-19 pandemic has begun to subside into endemic complacency even as new pathogens loom, the erosion of democracy continues at a newly accelerated pace. As long as so many unreflecting Robert Nevilles, dedicated to preserving a world already lost, surround us, exhort us, command us, and threaten us, and so long as we remain self-quarantined, either literally or metaphorically, with only incomplete and tentative knowledge, trust, and political will to guide us, we must acknowledge the possibility that their denialism may even reside within us.

Notes

1 See Zani and Meaux 98: "The most significant change in zombie cinema occurred in 1968, with George Romero's *Night of the Living Dead*, loosely based on Richard Matheson's 1954 vampire novel *I Am Legend*. Consequently, it is fair to argue that the blueprint for modern zombie cinema is Matheson's novel." See also Christie 67–68.

2 My analysis in this chapter is inspired by and to a certain extent modeled on Steven Shaviro's account of Romero's original zombie film trilogy as an allegory of contemporary capitalism, "Contagious Allegories" (Shaviro). Readers should also be aware that my analysis will make an important detour through the history of racist tropes in US culture.

3 All further unspecified parenthetical references will be to Matheson's novel.

4 For a thorough explication of *I Am Legend*'s symbolic economy of (anti-)Blackness that is highlighted in this passage, see Patterson. Although most fiction writers working with the new materials that Cold War paranoia had provided to the apocalyptic imagination focus either on the threat of nuclear annihilation (e.g., Nevil Shute's *On the Beach* [1957], Peter George's *Red Alert* [1958], and Mordecai Roshwald's *Level 7* [1959]) or on the threat of communist infiltration (e.g., Robert Heinlein's *The Puppet Masters* [1951] and John Wyndham's *The Midwich Cuckoos* [1957]), Matheson's incorporation of references to racial segregation, civil rights struggles, and the arguments around them is not entirely unique in apocalyptic fantastika of this period. Despite the fact that it is usually assimilated to the communist infiltration subgenre, Jack Finney's *The Body Snatchers* (1955, like Matheson's book the source of three popular film adaptations, in 1956, 1978, and 2007) also includes a very striking passage (Finney 137–39) that likens

the ironic mockery audible in the voices of the pod-spawned replacement people to that expressed by an African American shoe-shiner's verbal mockery of his own submissive banter with his white clients. Finney's protagonist-narrator Dr. Bennell sums up that tone as follows: "never before in my life had I heard such ugly, bitter, and vicious contempt in a voice, contempt for the people taken in by his daily antics, but even more for himself, the man who supplied the servility they bought from him" (Finney 139). On this passage in *The Body Snatchers*, see Wald 43–47.

5 In this regard, Matheson's novel also stages and implicitly criticizes some of the key features of reverse colonization narratives which, as David M. Higgins (2021) has shown, have been taken up by alt-right movements, in the form of fantasies such as the "Great Replacement" of white Americans by people of color or the Russian assertion of a right to "de-Nazify" Ukraine in order to protect itself, to justify pre-emptive reactionary violence as well as anti-democratic policies.

6 See Payne 9: "As well as the deliberate removal of a veil, the Greek word [*apokalypsis*] names the unfolding of a flower from its protective sheath, as what had been hidden from view emerges into sight through a nonvolitional process."

Works Cited

Calhoun, John C. *A Disquisition on Government and Selections from the Discourse*, ed. C. Gordon Post. Hackett, 1953. First published 1853.

Christie, Deborah. "A Dead New World: Richard Matheson and the Modern Zombie." *Better Off Dead: The Evolution of the Zombie as Post-Human*, ed. Deborah Christie and Sarah Juliet Lauro. Fordham UP, 2011, pp. 67–89.

Constitution of the United States of America. US Government Printing Office, 1992.

Douglass, Frederick. *Narrative of the Life of Frederick Douglass, an American Slave*. Project Gutenberg, 28 February 2021, https://www.gutenberg.org/cache/epub/23/pg23-images.html. Accessed 7 June 2023. First published 1845.

Finney, Jack. *Invasion of the Body Snatchers*. Orion, 2010. First published 1955.

Fukuyama, Francis. *The End of History and the Last Man*. Free Press, 1992.

Higgins, David M. *Reverse Colonization: Science Fiction, Imperial Fantasy, and Alt-Victimhood*. U of Iowa P, 2021.

Johnson, James Weldon. *The Autobiography of an Ex-Colored Man*, ed. Jacqueline Goldsby. W.W. Norton, 2015. First published 1912.

La Rochefoucauld. *Oeuvres* vol.1, nouvelle edition, ed. D.L. Gilbert and J. Gourdault. Hachette, 1923. First published 1665.

Matheson, Richard. *I Am Legend*. Tor, 1995. First published 1954.

Milton, John. *Paradise Lost*. *Poetical Works of John Milton*, ed. Rev. H.C. Beeching. Project Gutenberg, 1 May 1999, https://www.gutenberg.org/cache/epub/1745/pg1745-images.html#link2H_PREF. Accessed 7 January 2023. First published 1667.

Nixon, Rob. *Slow Violence and the Environmentalism of the Poor*. Harvard UP, 2011.

Patterson, Kathy Davis. "Echoes of *Dracula*: Racial Politics and the Failure of Segregated Spaces in Richard Matheson's *I Am Legend*." *Journal of Dracula Studies* 7 (2005), pp. 19–27.

Payne, Mark. *Flowers of Time: On Postapocalyptic Fiction*. Princeton UP, 2020.

Shaviro, Steven. *The Cinematic Body*. U of Minnesota P, 1993.

Shelley, Mary. *Frankenstein, or, The Modern Prometheus*. Project Gutenberg, 13 March 2013, https://www.gutenberg.org/cache/epub/42324/pg42324-images.html. Accessed 29 August 2023. First published 1831.

Twain, Mark. *The Adventures of Huckleberry Finn.* Project Gutenberg, 29 June 2004, https://www.gutenberg.org/cache/epub/76/pg76-images.html. Accessed 7 January 2023. First published 1884.

Wald, Priscilla. "Viral Cultures: Microbes and Politics in the Cold War." *Zombie Theory: A Reader*, ed. Sarah Juliet Lauro. U of Minnesota P, 2017, pp. 33–62.

Zani, Steven, and Kevin Meaux. "Lucio Fulci and the Decaying Definition of Zombie Narratives." *Better Off Dead: The Evolution of the Zombie as Post-Human,* ed. Deborah Christie and Sarah Juliet Lauro. Fordham UP, 2011, pp. 98–115.

Part 2

Intersectional Critique

4 Fungal Imaginaries

The Reconfiguration of Post-Pandemic Society in *Severance* and *The Last of Us*

Matthew Leggatt

In the 1980s, Giles Deleuze and Felix Guattari, in their book *A Thousand Plateaus*, used the rhizome as allegory to reimagine semiotics and epistemology. For the philosophers, the rhizome illustrated "principles of connection and heterogeneity," symbolizing an invisible system in which "any point … can be connected to anything other, and must be" (7). This was a progressive, horizontal model that would challenge the existing vertiginous modes of capital accumulation and knowledge. As that decade advanced, natural paradigms were becoming irresistible as metaphors for an emerging global system advanced by the spread of a dominant neoliberalism in the economies of North America and Europe in particular. For enthusiasts of globalization in the 1990s, who ultimately dominated in neoliberal political circles, Anna Tsing highlights that capital flows were regarded as life-sustaining and "world-making" ("The Global Situation" 327). But, this was far from the progressive model envisioned by Deleuze and Guattari, and many economic and cultural critics resisted the seduction of this natural imagery. See, for instance, David Harvey's work on time-space compression and the postmodern condition, or Scott Lash and John Urry's influential ideas about "disorganized capitalism," in which natural metaphors prove not only flawed and inadequate, but actively harmful. For Tsing, writing in 2000, ideas about circulation had come to displace Marxist critiques of capital penetration and the difference was telling: "where penetration always evokes a kind of rape," she writes, "a forcing of some people's powerful interests onto other people, circulation calls forth images of the healthy flow of blood in the body and the stimulating, even-handed exchange of the marketplace" ("The Global Situation" 336). By the mid-2010s, however, Tsing herself was returning to the rhizome to explore ideas about capital flows, albeit with an interest in capital ruination (i.e., what happens when flows stop?). In her book *The Mushroom at the End of the World*, she writes that "the uncontrolled lives of mushrooms are a gift—a guide—when the controlled world we thought we had fails" (2). As an organism that thrives in decaying spaces, Tsing adopts the mushroom as a much-needed image of hope for the twenty-first century: "Global landscapes today are strewn with [the ruin associated with global capitalism]," she writes. "Still, these places can be lively despite announcements of their death …. In a global state of precarity, we don't have choices other than looking for life in this ruin" (6). For many, the experience of the COVID-19 pandemic, which resulted

DOI: 10.4324/9781003345770-7

in the first of a number of national lockdowns in the UK in April of 2020, mirrors precisely this sense of the controlled world we thought we had failing, and our collective response, of which this collection is a part, reflects our scrabble to make meaning from this failure.

In Ling Ma's 2018 novel, *Severance*, published before the COVID-19 virus emerged but certainly sharing some uncanny resonances, a fungal virus quickly leads to the end of global civilization. In New York City, Candace Chen, the daughter of Chinese immigrants, clings to her unfulfilling job at Spectra, a company that oversees the production of elaborate versions of the Bible. In *Severance*, the effects of the fungal virus, known as Shen Fever, are often difficult to tell apart from ordinary human behavior. The implication is clear that the state of zombified nostalgia afflicting the infected is not dissimilar from feelings experienced by Candace and a number of other characters. While there is certainly a debate as to what meaning might be drawn from the disaster, Candace is minded that "the fever was arbitrary. The fact that [Candace and her group] were alive held no special meaning" (32). Suggestions that the pandemic and its associated outcomes are ultimately meaningless because they are driven by chance (chance encounters with the virus, chance susceptibility or immunity, chance viral mutations, etc.) place us as bystanders in, and victims of, a global system over which we can assert no control and thus in which we might as well have no stake. However, in this chapter, I want to offer a counterview of the pandemic which moves beyond the suggestion that our contemporary moment is characterized simply by disorganization and suggest that, while the events of the pandemic may sometimes appear disorganized, they are, in fact, organically structured much like the rhizome. As I will show, at times, this underlying, horizontal structure creates a dangerous appearance of randomness that, if harnessed by malevolent forces, can be extremely damaging and can mask the logic of embedded systems of inequality. At the same time, however, and more in the spirit of Deleuze and Guattari, these fungal structures have already proven vital in our efforts to combat COVID-19, making us more resilient as a species and helping to connect us in ways that may eventually be transformational.

Lash and Urry argued in the late 1980s that the world was entering a new phase in global economic relations they termed "disorganized capitalism." It was characterized by, among other things, deregulation, particularly in the financial sector; an expansion in white collar or service-based jobs also resulting in educational changes, and the emergence of new social movements. Lash and Urry also note the shrinking importance of the state as a disorganizing factor since state power could previously be relied upon to provide a unified response to global affairs but was now being dwarfed by the power of transnational corporations and other actors on the global stage (6). These trends, I argue, have seen something of a reversal post-COVID-19. For example, where Urry and Lash suggested that new political social movements "increasingly draw energy and personnel away from class politics," (5) a number of factors (most notably inflation, increasing interest rates, long-term stagnation of wages, austerity, increased unionization and strike action, various political scandals, and now even advancements in artificial intelligence) have been driving a renewed sense of class consciousness in recent years. As John Preston and Rhiannon Firth argue in an early response to the COVID-19 pandemic: "capitalism creates and maintains capital at a multitude of scales, including the viral" and governmental responses to COVID-19 often took the form of continuing class warfare (4). Thus, increased state interference into everyday lives through national lockdowns, vaccine mandates, and constantly shifting COVID-19 policies has worked to push state power back to the forefront of global politics, at least temporarily, and has also seen a resurgence in class-based politics, albeit in a sense complicated by disinformation spread by mainstream political parties. Thus, while I've already noted above that one response to the pandemic might be to see events as arbitrary, meaningless, or as further evidence of the disorganization of our contemporary moment, this tends to ignore new sources of connection and unification that also emerged out of the pandemic.

I prefer to return to the rhizome to articulate the growth and interdependence of this, seemingly illusive, network of economic nodes. Today, as we emerge slowly, and stutteringly, from the shadow of COVID, the rhizome offers a model to help us understand the orientation of our society and the transformation of our world. In some senses, these fungal economies facilitate corruption because fungal capital is not really that arbitrary at all: the connections between nodes may not always be immediately visible, but they do exist below the surface. They are there to be uncovered by following the money, as with the PPE (personal protective equipment) scandal where members of the UK government oversaw the handing out of lucrative contracts to some of its wealthiest friends and donors to manufacture masks and other protective equipment (see Conn and Lewis on the Michelle Mone scandal). Despite this potential for exploitation in ways that might be difficult to anticipate, such fungal networks can also be utopian since, as Deleuze and Guattari indicate, rhizomes offer a different model to the top-down hierarchies of neoliberal economics and, therefore, provide the possibility of resistance. Connecting us, as they do, far and wide, in often interesting and unexpected ways, the rhizomic networks of the post-pandemic world have led to the formation of new fungal relationships and communities; new fungal ways of working and socializing; and new fungal solutions to problems involving collaboration. This is, perhaps, most evident in changes to working patterns that have arisen out of COVID-19 lockdowns and restrictions. While during the pandemic these restrictions meant isolation for many, as freedoms returned the supportive networks built to sustain us in lockdown remain for many. It is not that technologies like video and conferencing calling didn't exist prior to the pandemic, but simply that they were used by relatively few in the workplace where now such technologies have become ubiquitous. This opens up real possibilities for stronger collaboration between researchers and activists across the world, it connects people from different workspaces all over the country and the globe, it means less frequent travel to work and so could have positive impacts on traffic and vehicle use. Beyond this broadening of networks, forced separation has also made many re-evaluate their interpersonal relationships with family and friends, strengthening them and helping us to see the value of keeping in touch and in-person socialization.

In this chapter, I explore HBO's *The Last of Us* (2023) and Ling Ma's *Severance*, both of which feature a fungal-born zombie infection that wipes out the recognizable structures of North American late capitalism. I argue that these quite differing perspectives on fungal imaginaries of post-pandemic worlds might help us to rethink the value of the connections and networks that have emerged in the wake of the COVID-19 outbreak, both in terms of their potential to help develop constructive relationships with others at home and abroad and in terms of the dangers presented if these networks are exploited by those who seek to sow confusion and distrust.

The Last of Us

In a notable addition to its videogame originator, *The Last of Us*, a 2023 television series created by Neal Druckmann and Craig Mazin and starring Pedro Pascal and Bella Ramsey, opens with a segment from a fabricated talk show set in 1968. In the set piece, two epidemiologists, Dr. Schoenheit and Dr. Newman, square off over what each perceives as the most significant threat to the planet. Given the context for the viewer, with the world still emerging from the COVID-19 pandemic, we are immediately confronted with what appears to be a prophetic narrative as Schoenheit describes a pandemic in which everyone gets sick all at once, based on a new influenza-type virus that is airborne and that, thanks to modern travel (air-travel most notably), could move from Madagascar to Chicago within a week. The audience, however, is more likely to be interested in the follow up from Dr. Newman, who claims not to be worried about such a virus: "mankind has been at war with the virus from the start. Sometimes millions of people die as in an actual war but, in the end, we always win," he insists. Instead, Newman warns, it is the fungus that keeps *him* up at night.

Put in its appropriate context, this is a rather playful opening. One might, entirely reasonably, find the pandemic described by Schoenheit terrifying, but the viewer will inevitably be drawn in by Newman's intriguing concerns about the dangers of fungi. When Newman is challenged by Schoenheit on the grounds that fungi cannot infect humans in the same way they infect other creatures, Newman counters that this has to do with body temperature and insists it could all change: "what if, for instance, the world were to get slightly warmer," he asks. Not content to blame the killer fungus to come on careless human activity, Newman then goes on to describe something that we might read in more abstract terms. He lists several genera of fungus, suggesting that

> any one of them could become capable of burrowing into our brains and taking control not of millions of us but billions of us; billions of puppets with poisoned minds, permanently fixed on one unifying goal: to spread the infection to every last human alive by any means necessary.

In this final address, he obviously anticipates the fungal infection the show will center around, but his language is also provocative. That the fungus is described as capable of "burrowing into our brains" and "poisoning" our minds aligns it, I would argue, with another fungal phenomenon to hit us in our pandemic moment: the rise and spread of disinformation in the wake of the COVID-19 pandemic and, in particular, vaccine conspiracy theories that ultimately "unify" us to help spread the plague among ourselves. Once again, human action or, more specifically, inaction is to blame.

Teija Sederholm, Petri Jääskeläinen, and Aki-Mauri Huhtinen also note the rhizomatic nature of the spread of disinformation that followed in the wake of the COVID-19 pandemic. Facilitated by the uncertainty surrounding the virus and the sheer variety of local and global actors who queued up to have their say online, they note that the resultant conspiracy theories were also aided by the fact they were generally "not tied to any particular political spectrum or geographical location," making them "open for new connections supporting [their] basic claim[s]" (44). In short, where new technologies played a vital role during the COVID-19 pandemic as components of an emerging, digital infrastructure that would allow many of us to keep working and stay in touch in ways that simply would not have been possible even a decade earlier, some of these same technologies, particularly those that generally work against hierarchies of knowledge, were also responsible for misdirection and miscommunication. Rather than flowing along neatly carved channels, the massive amounts of information the pandemic generated formed fungal networks that were, and still are, loosely connected but also rather unpredictable.

As with the oblique reference to COVID-19 in the opening episode of *The Last of Us*, episode two also starts with a return to the pre-fungal world—in fact, to the origin of the outbreak in Jakarta, Indonesia, where a professor of mycology, Ratna Pertiwi (Christine Hakim), has her lunch interrupted by military personnel who take her to examine the body of one of the first known victims of a deadly fungal infection. In this opening, a world unfamiliar to the Anglophone viewer passes by but, as the professor arrives at the facility, one thing stands out: a recognizable word, SARS. It appears on one of several standing banners in the facility past which the professor is escorted and serves to remind the audience that COVID-19 is not, in fact, the first such viral outbreak (nor will it be the last—as Dr. Newman suggests, we have always been at war with the virus). SARS (Severe Acute Respiratory Syndrome) was the name given to the disease caused by a coronavirus first discovered in China in November 2002. In its recollection of SARS, *The Last of Us* calls attention to what David Quammen has called a "lucky" escape for humanity (207). Despite the fact that in the era of globalization, "everything nowadays moves around the planet faster, including viruses," with SARS those who were infected tended to become most infectious *after* symptoms had set in which, unlike what we have seen with COVID-19, helped authorities contain the spread (207).

In fact, where the origin of the fungal outbreak in the game is 2013 (the same year that the game was released), Druckmann and Mazin adapted this for the series in which the outbreak instead occurs a decade earlier in 2003. In an interview for *Insider*, the creators claim that the motivation for the change was to "allow the show's main timeline [where the action takes place 20 years after the outbreak] to be now, in 2023" (Acuna). Rather than SARS, the creators suggested that they were inspired by the 1918 Spanish flu pandemic and, particularly, "how people reacted and how they became, in some ways, xenophobic and kind of closed off their tribes." According to Kirsten Acuna, "Druckmann added that the intention of the HBO adaptation wasn't to focus on the current pandemic" and this might explain the subtlety of the hints the creators make:

> It became important to us actually not to comment on the modern outbreak too much because we didn't wanna make a Covid 19 show .... Ultimately, this is a show that's post-pandemic. The pandemic itself is a part of the show, and then we deal with 20 years after that .... How are people dealing with this new reality? (Neil Druckmann, cited in Acuna)

By setting the outbreak in 2003, however, the creators recall not just the xenophobia associated with the Spanish flu but also that unleashed by SARS-CoV-1.

A significant number of studies track anti-Chinese racism after SARS in 2003 and note the way in which the virus became racialized—Carrianne Leung (2008) and Katherine A. Mason (2015), in particular. By comparing the response to SARS in 2003 and the H1N1 (swine flu) pandemic in 2009, Mason shows the complex way in which such racialized rhetoric develops in different regions: "While SARS in this imaginary became a 'Chinese' or 'Cantonese' disease," she argues, "H1N1 became a EuroAmerican disease that, when it reached inside China, adhered more easily to those Chinese who did not quite belong" (500). In this context, the brief reference to SARS in *The Last of Us* recalls, uncomfortably, President Trump's repeated labelling of COVID-19 as the "China" or "Chinese Virus" that, according to a study by Hswen et al. (2021), was connected to a significant rise in anti-Chinese and anti-Asian "hate expressions" on *Twitter* (962). Once again, we see the rhizomic way in which information spread online in the wake of such an event takes on both intended and unintended consequences. Where online hate speech is often spoken about as viral in nature, virality tends towards linearity and predictability. Rhizomic behaviors, by contrast, tend towards unpredictability. I prefer this way of thinking about the spread of information in the wake of COVID-19. Trump was, by this point, known for making bombastic statements, and controversy was his Presidential hallmark. However, while propaganda allows for the careful curation of a political message, misinformation, or in this case disinformation, is in many ways its opposite since in the environment created by the pandemic and still evident since, where online conspiracy is rife, the interpretation of such information can be near impossible to control in any predictable way.

Zombies have become particularly ambiguous archetypes in terms of their symbolic resonance. Pielak and Cohen, for instance, remark that "zombies reflect the culture that creates them" (2). While in the twentieth and twenty-first centuries, particularly during times of economic crisis, zombies have often been regarded as symbolic of both capitalism and consumerism, with David McNally noting in the wake of the 2008 financial crisis references to "zombie banks," "zombie economics," "zombie capitalism," and "zombie politics" in the media (1), zombies have also been explored fruitfully in relation to the Other (cf. the recent collection *Decolonizing the Undead: Rethinking Zombies in World-Literature, Film, and*

Media and Camilla Fojas' book *Zombies, Migrants, and Queers: Race and Crisis Capitalism in Pop Culture*, among others). In this sense, they can offer "a more radical reimagining of the possibilities of living and living together" (Pielak and Cohen 9). One might, quite legitimately then, ask what, in *The Last of Us*, the zombified masses represent? Are they climate change deniers? Vaccine conspiracy theorists? Those whose minds have been "poisoned" by the Rupert Murdoch news outlets, becoming puppets for an emerging fascism whose proponents seek to weaponize disinformation? With the casting of Chilean-born Pascal and the non-binary Ramsey in the lead roles, alongside an entire episode being dedicated to a romantic subplot between two male characters, the show does seem an overt rejection of the politics of Fox News. But, the show also seems cynical about human nature. For example, consider the wretchedness of the survivors in *The Last of Us*—a show where even the hero, Joel, is a cold-blooded killer who selfishly murders humanity's hope for a cure in the final scenes in order to save his daughter surrogate, Ellie—or the events of episode five in which the protagonists themselves need to be rescued *by* the fungal zombie hoard from humans who are even more violent and driven by vengeance than they are. Perhaps, the show's ambiguity here is precisely the point. It serves to highlight that there are certainly positive outcomes that emerge from connection and collaboration, but that networks can also become infected, or poisoned, turning us against one another. In the sweeping and gestural politics of *The Last of Us*, connection is both something to be feared and yet also, paradoxically, the only way to survive.

Severance

In Ling Ma's *Severance*, a text that Ivan Franceschini has suggested "is really a metaphor for the alienation of labor in contemporary capitalist society," we see once again fungal zombies that are somewhat ambiguously treated. One telling scene in particular demands attention (245). As the novel approaches its anticlimactic close, protagonist Candace Chen observes a fungal zombie as it continues with its daily routine, folding and refolding the same shirt over and over even as half of its jaw is missing. A powerful image of the sweatshop sublime is conjured in this moment: a call to think of our place in the global economy, which is normally met with a type of pacification—perhaps, zombification—is, at the end of this novel, instead rendered powerful and transformative, even more so when re-read in the wake of the COVID-19 pandemic.

The *sweatshop sublime* is a term coined by Bruce Robbins in his 2002 journal article of that very name. In the article, Robbins uses incidents from two texts to illustrate the sublime encounter with the global machine and its pacifying effect on us. Referring to a passage in David Lodge's 1988 novel, *Nice Work*, Robbins writes:

> To contemplate one's kettle and suddenly realize, first, that one is the beneficiary of an unimaginably vast and complex social whole and, second ..., that this means benefiting from the daily labor of the kettle- and electricity-producing workers, much of it unpleasant and underremunerated, is not

> entirely outside everyday experience. What seems special about the passage is a third realization: that this moment of consciousness will not be converted into action.
>
> (84)

In another moment, Robbins describes someone donning a t-shirt in the morning and looking at the label. In an instant they are transported "from the tender, drowsy privacy of early morning at home […] to the outer reaches of a world economic system of notoriously inconceivable magnitude and interdependence, a system that brings goods from the ends of the earth … to satisfy your slightest desire" (85). Yet, according to Robbins, "this insight is strangely powerless," and, without much further thought, the individual is likely to simply get dressed and carry on with their day, putting the uncomfortable encounter to the back of their mind (85).

There is something telling in this account, particularly in our present state. Through the image of the sweatshop worker, the novel demands that we empathize with the zombies of exploitative capitalist labor. But, in this post-pandemic moment, we might also stop to think about our connection to those around the world: both those that suffer and those who come together. For example, we might ask why the amazing achievements of vaccine scientists, able to work together across the globe to understand COVID-19 and prevent its most severe symptoms and side-effects, might not fill us with a sense of empowerment; we might think about the ability of governments (particularly in rich countries) to roll out huge vaccination programs; we might applaud and celebrate the incredible efforts of the doctors, nurses, and paramedics who worked to battle the virus when hospitals were overflowing with patients. Why should the COVID-19 pandemic produce in us that same sensation of powerlessness associated with the sweatshop sublime? Perhaps, because, as many of us became siloed in our homes, we felt helpless: at the mercy of the whims of both nature and the enigmatic global economic and political systems over which we seem to have no control.

Ultimately, where *The Last of Us*, by its nature, offers us a story about human agency (based as it is on a first-person shooter videogame in which you control individuals who will eventually fight their way through the zombie horde towards a particular goal), *Severance* is a text about its opposite: Candace's lack of agency. Both after the pandemic, where, like the fevered, she continues to work her day job even as the world around her vanishes and before—Chapter 3 opens with "I arrived to the city carried by the tides of others. … It seemed like the inevitable, default place to go" (34)—Candace's story is characterized by aimlessness. This is, perhaps, a product of the novel's strategy to reveal the impacts of neoliberalism (we might note, on a surface level, Ma's choice to name the novel *Severance*). The hopelessness experienced by Candace seems to offer a response to a world in which, despite the obvious collapse of the global economic system, most people carry on as they are. Emrah Atasoy and Thomas Horan note that "though the effects of neoliberalism, as this literature illustrates, are painful and at times deadly on a mass scale, its causal role in suffering often goes unrecognized" (Atasoy and Horan 237). They quote Eugene McCarraher, who rather fatalistically declares, "Even in the face of its manifest injustice, degradation, and ecological toxicity, capitalism remains for most Americans the horizon of moral and political possibility" (McCarraher 669). This helps to explain the somewhat unsatisfying ending of Ma's novel, which Burcu Akkoyun has argued suggests that "to live in the city … is actually to become fevered with what [Zygmunt] Bauman identifies as 'the dreary uniformity of the urban habitat'" (291). Thus, for the reader of *Severance*, while prior to the collapse of the global economic system we encounter a global economy set up to exploit and extract, the fungal potentiality released by Shen Fever does not entirely liberate Candace from participation in hierarchies of power and exploitation.

In other ways, however, the pessimism evident in *Severance* seems to tally with what has already been noted about *The Last of Us*. At points, the novel's narrator ruminates on the chaotic nature of the information shared about the fungal infection in the days and weeks after its discovery, and, in particular, the inability to discern fact from fiction:

> Who knew what was true. The sheer density of information and misinformation at the End, encapsulated in news articles and message-board theories and clickbait traps that had propagated hysterically through retweets and shares, had effectively rendered us more ignorant, more helpless, more innocent in our stupidity. (31)

As in *The Last of Us*, where the spread of the fungus renders humans *poisoned puppets*, in *Severance*, the infected revert to a form of innocent stupidity. Their experience of music is illustrative since they fail to register "the emotional content" of the songs they hear but, nevertheless, respond to rhythms and "percussive patterns that had worn grooves inside their brains" (28). It is, perhaps, telling that the novel deals not with zombies in the traditional sense of the word—Candace, herself, points out that, "the fevered aren't zombies. They don't attack us or try to eat us. They don't do anything to us. If anything, we do more harm to them" (29)—but with another collection of bodies that have been subdued and incapacitated, their brains emptied of whatever permits individual thought. Indeed, for a time, some of those who are not infected choose to live alongside loved ones, still naively hoping for a cure to be discovered, and, like Candace, unable to let go of a life that is disappearing slowly before their very eyes.

For Funke, "The image of the rhizome … evokes something that is always in the process of becoming and of growing in unplanned ways" (360). He highlights that, for Deleuze and Guattari, "[a] rhizome doesn't begin and doesn't end, but is always in the middle, between things, interbeing, intermezzo" (Funke 360). This in-betweenness is evident in Candace's uncertain identity. With her Chinese parents having passed away, the novel sees her caught precariously between a desire to connect with her Asian heritage and a sense of youthful New Yorker indifference. When her friend incorrectly assumes she'd know how to play mah-jongg, she retorts simply "'Why, because I'm Asian?'" (53) but in recalling her mother's skin care regime which she witnessed as a child she laments that she had looked for a particular Fujianese face oil that her mother had used "everywhere, in both

Cantonese China-towns, in Fujianese Chinatown, in Sunset Park, in Flushing, and never found it" (63). Serrano-Muñoz notes that "[*Severance*] is also an honest and complex exploration of migrant and minority identities within global capitalism" (1353). In this way, Ma's novel exposes, or at least hints at, the complex mesh of interconnected threads that make up the global rhizomatic network of our times:

> Candace becomes … representative of a new model of intersectional identity that speaks of multiple crises: young, politically critical but not an activist, an independent female character, an uprooted migrant that is not traumatized by nostalgia but adapted—albeit not painlessly—to fluidity. Candace is defined by the combination of all of these circumstances; the novel does not hierarchize the fact that she is a woman, or an Asian American, or a precariat professional as isolated conflicts. Instead, this protagonist reinforces the message that crises need to be understood as systemic entanglements that must be approached and interpreted together in order to see the whole picture. (Serrano-Muñoz 1354)

Thus, perhaps in its multiplicities, *Severance* does offer a larger overarching message about our times. Again, however, we must look to the fungus to help reveal it.

Citing Deleuze and Guattari, Funk highlights that the rhizome suggests an "unplanned," spontaneous, or at least under-strategized direction. By its very nature, the rhizome lacks "genetic axis or deep structures" (360). Thus, fungal connections offer the antithesis of the horizontal and embedded structures associated with an entrenched and resilient modern capitalism. But these structures in capitalism do not necessarily pertain to relationships between people who the system, instead, conditions us to see in a utilitarian capacity. This is embodied in *Severance* by the fleeting and chance encounters Candace has with various men when she moves to New York—a striking contrast to the more meaningful and deep connection between Joel and Ellie that drives the narrative of *The Last of Us*. Candace states,

> All I did that first summer in New York was wander through lower Manhattan, wearing my mother's eighties Contempo Casuals dresses, looking to get picked up by anyone, whomever. The dresses slid on easily in the morning. They slid off easily at night. They were loose-fitting and cool, cut from jersey cotton in prints of florals and Africana. Wearing them, I never failed to get picked up but I usually failed to get anything else—not that I wanted anything else, as I told myself and whomever else. Still, I overstayed my welcome in their beds, wondering what they did for a living as they dressed in the mornings. Where they were going. (34–35)

Where Candace claims a certain distance from the men she sleeps with, there is also evidently a deeper lack of satisfaction that haunts this image of a young woman unable to connect once morning arrives. Like the t-shirt in Robbins' example of the sweatshop sublime, the fashion statement dresses that Candace wears slip on and off like they do nothing more than mark the time. These dresses connect her to the global economic system, with their floral prints and Africana stylings, but the connection is superficial, like tenuous fungal mycelia that sit just below the surface of the soil. Thus, through its microscopic politics of passing connections, *Severance* points to a larger truth about the post-pandemic rhizome: while an incredible network has, at least briefly, united us, opportunities for deeper and more meaningful exchanges—the kinds of exchanges and collaborations that are so desperately needed in a world teetering on the brink—may be passed up if we return to our siloed lives of fleeting encounters in the night.

Concluding Remarks

According to Tsing, "spores open our imaginations to another cosmopolitan topology. Spores take off toward unknown destinations, mate across types, and, at least occasionally, give rise to new organisms—a beginning for new kinds" (*Mushroom* 277). It is this rather refreshing and quietly optimistic view of fungal potentiality on which I wish to end. As I have tried to demonstrate, both *Severance* and *The Last of Us* offer different visions as part of their fungal imaginary. *The Last of Us* gives us the macro-politics of the pandemic by way of its sweeping, gestural politics and its exploration of disinformation warfare, unity, climate change and responsibility, and familial love and human nature. By contrast, *Severance* deals with the disjunctive, tangential, seemingly arbitrary micropolitics that suggests a certain hopelessness in our predicament. To this extent, both offer us different visions of our present rhizomic relations to the global whole. One, perhaps, more positive than the other, but both existing simultaneously and exhibiting an element of truth about what might seem otherwise to be an arbitrary connectedness. These have certainly been difficult times and we should be wary of new fungal relationships that might exist only tenuously or temporarily, and new parties that might exploit networks of disinformation in order to turn us into poisoned and brainless zombies, but, as Tsing puts it, might this also be "a beginning for new kinds"? We should think of our place in this global whole, the way that so many collectively responded to the dangers of the pandemic through collaboration, togetherness, and scientific ingenuity, not with a sublime powerlessness, but with a sense of bright optimism that we can, collectively, do things better. Let us be hopeful that this fungal system can lay down deeper roots and bring about a new abundance of life in the decaying structures left behind by a dying system.

Works Cited

Acuna, Kirsten. "Why 'The Last of Us' Show Kicks Off a Decade Earlier Than the Video Game." *Insider*. www.insider.com/why-the-last-of-us-show-starts-2003-decade-earlier-2023-1. Accessed April 2023.

Akkoyun, Burcu Kayişci. "Urban Apocalypse, Global Precarity, and Uncanny Liminality in Colson Whitehead's *Zone One* and Ling Ma's *Severance*." *Critique: Studies in Contemporary Fiction*, vol. 64, no. 2, 2023, pp. 282–95.

Atasoy, Emrah and Thomas Horan. "Prayer Had Broken out: Pandemics, Capitalism, and Religious Extremism in Recent Apocalyptic Fiction." *Studies in the Novel*, vol. 54, no. 2, 2022, pp. 235–54.

Conn, David and Paul Lewis. "Revealed: The Full Inside Story of the Michelle Mone PPE Scandal." *The Guardian*, 9 December 2022. https://www.theguardian.com/uk-news/2022/dec/09/revealed-the-full-inside-story-of-the-michelle-mone-ppe-scandal.

Deleuze, Giles and Felix Guattari. *A Thousand Plateaus*. Translated by Brian Massumi, University of Minnesota Press, 1987.

Fojas, Camilla. *Zombies, Migrants, and Queers: Race and Crisis Capitalism in Pop Culture*. University of Illinois Press, 2017.

Franceschini, Ivan. Zombies of Capital: On Reading Ling Ma's Severance, *Made in China Journal*, 2020. https://madeinchinajournal.com/2020/06/10/zombies-of-capital-on-reading-ma-lings-severance/.

Funke, Peter Nikolaus. "The Global Social Forum Rhizome: A Theoretical Framework." *Globalizations*, vol. 9, no. 3, 2012, pp. 351–64.

Harvey, David. "Time-Space Compression and the Postmodern Condition." *The Condition of Modernity: An Enquiry into the Origins of Cultural Change*. Blackwell, 1989, pp. 284–307.

Hswen, Yulin, Xiang Xu, Anna Hing, Jared B. Hawkins, John S. Brownstein, and Gilbert C. Gee. "Association of "#covid19" versus "#chinesevirus" with Anti-Asian Sentiments on Twitter: March 9–23, 2020." *American Journal of Public Health*, vol. 111, no. 5, 2021, pp. 956–64.

Lash, Scott and John Urry. *The End of Organized Capitalism*. The University of Wisconsin Press, 1987.

The Last of Us, created by Neil Druckmann and Craig Mazin, HBO, 2023.

Leung, Carrianne. "The Yellow Peril Revisited: The Impact of SARS on Chinese and Southeast Asian Communities." *Resources for Feminist Research*, vol. 33, no.1, 2008, pp. 135–49.

Ma, Ling. *Severance*. The Text Publishing Company, 2020.

Mason, Katherine A. "H1N1 Is Not a Chinese Virus: The Racialization of People and Viruses in Post-SARS China." *Studies in Comparative International Development*, vol. 50, 2015, pp. 500–18.

McCarraher, Eugene. *The Enchantments of Mammon: How Capitalism Became the Religion of Modernity*. Harvard University Press, 2019.

McNally, David. *Monsters of the Market: Zombies, Vampires and Global Capitalism*. Brill, 2011.

Pielak, Chase and Alexander H. Cohen. *Living with Zombies: Society in Apocalypse in Film, Literature and Other Media*. McFarland and Company, 2017.

Preston, John and Rhiannon Firth. *Coronavirus, Class, and Mutual Aid in the United Kingdom*. Springer, 2020.

Quammen, David. *Spillover: Animal Infections and the Next Human Pandemic*. Vintage, 2020.

Robbins, Bruce. "The Sweatshop Sublime." *PMLA*, vol. 117, no. 1, 2002, pp. 84–97.

Sederholm, Teija, Petri Jääskeläinen, and Aki-Mauri Huhtinen. "Coronavirus as a Rhizome: The Pandemic of Disinformation." *International Journal of Cyber Warfare and Terrorism*, vol. 11, no. 2, 2021, pp. 43–55.

Serrano-Muñoz, Jordi. "Closure in Dystopia: Projecting Memories of the End of Crises in Speculative Fiction." *Memory Studies*, vol. 14, no. 6, 2021, pp. 1347–61.

Shapiro, Stephen, Giulia Champion and Roxanne Douglas (Eds.). *Decolonizing the Undead: Rethinking Zombies in World-Literature, Film, and Media*. Bloomsbury, 2022.

Tsing, Anna. "The Global Situation." *Cultural Anthropology*, vol. 15, no. 3, 2000, pp. 327–60.

Tsing, Anna. *The Mushroom at the End of the World: On the Possibility of Life in Capitalist Ruins*. Princeton University Press, 2015.

5 Five Hundred Years of Plague

Indigenous Apocalypse in Joca Reiners Terron's *Death and the Meteor*

Benjamin Burt

Introduction: Post-apocalyptic People

Epidemic disease has wreaked havoc on Indigenous peoples in the Brazilian territory since European contact. Whether transmitted unintentionally or weaponized by settlers eager to expand further inland, pathogens devastated native communities throughout the Americas. After 150 years of European colonialism, the total Amerindian population decreased by approximately 95% (Danowski and Viveiros de Castro 104). This demographic collapse inspired Indigenous Brazilian intellectual Ailton Krenak's claim that "the world ended in the sixteenth century" for many native communities (*Ideas* 69). Scholars Déborah Danowski and Eduardo Viveiros de Castro concur: "for the native people of the Americas, the end of the world already happened—five centuries ago. To be exact, it began on October 12, 1492" (104).[1] These quotations reveal the devastating impact of disease in the colonial period and hint at the subjective barrier between epidemic and pandemic. From a European perspective, the outbreaks sparking mass death were geographically isolated phenomena. For Amerindians, these plagues attained pandemic, world-annihilating proportions.

The descendants of those who survived this initial wave of destruction, including present-day communities, persist in an extended, post-apocalyptic historical period. In contemporary Brazil, Indigenous people comprise a mere 1% of the total population and face unyielding social marginalization. Despite making gains in recent decades, a lack of Indigenous representation in the political, artistic, and intellectual spheres foments a misguided belief that native cultures are a remnant of the historical past. Epidemic outbreaks may also recall the colonial period, but they remain a constant threat. Once-isolated tribes continued to fall victim to diseases transported by missionaries and other migrants throughout the twentieth century (Vilaça). The COVID-19 pandemic likewise took an outsized toll on native communities (Brum). Deprived of adequate official representation and besieged by disease and illegal enterprises like wildcat mining and illicit ranching, native communities must engage in constant resistance.

Having persevered through centuries of recurrent pandemics and violent invasion, contemporary Indigenous communities possess knowledge that may prove useful on a planet under threat from climate breakdown. Danowski and Viveiros de

DOI: 10.4324/9781003345770-8

Castro call twenty-first-century Amerindians "Veritable end-of-the-world experts" who are well-versed in strategies of adaptation to devastation (108). At the same time, however, "it would be no surprise if they were not interested in 'negotiating' any cosmopolitical peace, and deservedly let us [non-Indigenous people] burn in hell" (Danowski and Viveiros de Castro 123). As the Capitalocene grows increasingly post-apocalyptic, these long-persecuted peoples may play a key role in deciding humanity's fate.[2]

In his 2019 novel *Death and the Meteor* (*A morte e o meteoro*), Brazilian author Joca Reiners Terron draws from the plight of contemporary Indigenous groups to consider themes including cataclysm and post-apocalyptic renewal.[3] The brief yet intricately plotted novel's four chapters extrapolate the brutality of early twenty-first-century life into a dystopian near future where the Amazon rainforest has become "hotter than hell"[4] due to climate change (11).[5] On the edge of extermination after decades of epidemics and environmental degradation, the fictional Kaajapukugi tribe embodies the author's belief that "The end of the world is diachronic. It already happened. For the Indians [sic], it happened a long time ago" ("Joca Reiners Terron"). The Kaajapukugi represent a mélange of Amazonian cultures with parallels to specific groups, including the Yanomami, the Kanoê do Omarê,[6] and the lone survivor of a tribal massacre known only as "the Man of the Hole."[7] With no surviving female members, Terron's fictional group consists of 50 men whose continued reclusiveness initially suggests their resignation to eventual extinction.

Unexpectedly, the surviving Kaajapukugi set the plot in motion by requesting political asylum in Oaxaca, Mexico. Shortly after arrival, however, they consume the last of their sacred hallucinogen *tinsáanhán* and commit ritual suicide. The titular meteor arrives and destroys the Earth as the narrator, an unnamed Mexican anthropologist, transports the cadavers to Brazil by plane. The implication of this fantastical denouement is that the tribe's suicide somehow triggered the global apocalypse. As Terron reveals more of the Kaajapukugi's backstory, any desire for revenge against the outside world appears increasingly justified. However, the author complicates what originally appears to be a narrative of totalizing vengeance and instead crafts an unexpectedly nuanced view of cataclysm intertwined with genesis.

This chapter argues that *Death and the Meteor*'s unanticipated recall to hopefulness within its otherwise bleak social vision creates a productive tension that accentuates the limits of fatalism and the complexity of collective aspiration in post-apocalyptic scenarios. After presenting relevant plot summary, this chapter includes a brief theoretical section that incorporates pertinent information about apocalyptic narration and representations of Amerindian cultures. The subsequent textual analysis begins by juxtaposing the possibility that the meteor permanently ends life on Earth with the cyclical metaphysics of Kaajapukugi cosmology.[8] The following section examines this difference of perspective through the mythological figure of *Hen-zaogao*, the Great Evil. Next, the chapter delves deeper into Kaajapukugi temporality and the tribe's belief in eternal return. The conclusion examines the possibility of post-apocalyptic hope, asserting that Terron preserves an ambiguous sense of optimism despite his characters' belief in a predestined history full of suffering.

Terron narrates much of the Kaajapukugi's history through an intradiegetic recording made by the Brazilian anthropologist Boaventura. Now elderly, Boaventura lived with the tribe some decades prior and remains the leading authority on the reclusive group and its cosmology. However, he confesses his culpability for the tribe's demise in a long video sent to his Mexican colleague prior to his death. The Brazilian describes setting out for Kaajapukugi territory with good intentions, quarantining himself to limit the risk of "the dormant projectiles in my body that could decimate the whole population of that unknown people" (41). Nonetheless, he later becomes a vector of disease and catalyst of existential despair. After several months living in isolation at the edge of the Kaajapukugi settlement, Boaventura unexpectedly kidnaps the group's only female member. The young woman miscarries her pregnancy with a Kaajapukugi man, but later bears a son with Boaventura. Before succumbing to measles, she explains her tribe's cosmology of eternal recurrence to her captor. After the woman's death, Boaventura transports his infant offspring to Kaajapukugi land.

Arriving at the Kaajapukugi's sacred island with his son, Boaventura discovers an entombed corpse wearing a spacesuit. This figure recontextualizes the Chinese mission to colonize Mars that, until this point, appeared to be a colorful, futuristic detail with minimal connection to the primary narrative. Indeed, the narrator concludes that the ancient cadaver must belong to one of the taikonauts who were earlier lost in space. Terron hints at a connection between the Kaajapukugi and Asian cultures throughout the novel, including using pinyin Mandarin Chinese to stand in for the tribal language, yet this revelation represents a significant twist. Despite their contemporary manifestation as an isolated group, the fictional tribe's Chinese heritage creates a new association with one of most influential societies of the early twenty-first century. Further, Terron implies that the Kaajapukugi that descend from these Chinese progenitors are the common ancestors of all humankind. Through this disorienting shift of perspective, the author reveals the universality of an Indigenous group previously envisioned at the outmost margins of contemporary society.

Furthermore, the Kaajapukugi's metaphysics prove a surprisingly useful heuristic for interpreting a world in crisis. Although the narrator is initially skeptical, he ultimately comes to believe that the meteor's arrival represents both the end and beginning of an ever-repeating history bridged by the taikonauts. The reclusive, largely silent tribe embodies a universal knowledge that perseveres despite the tragic history of each metaphysical cycle. Whether this situation is one of totalizing despair or eternal hopefulness, however, remains open to interpretation.

Theoretical Overview: Assessing Apocalypse

The explosive ending of *Death and the Meteor* creates an obvious association with the popular understanding of apocalypse as the end of the world. Well before the conclusion, the dystopian setting and theme of impending extinction create points of connection with contemporary, post-apocalyptic fiction. At the same time, the final twist acts as a revelation that exemplifies the etymological meaning of apocalypse ("Apocalypse, n."). This apparent confirmation of the Kaajapukugi cosmology recontextualizes the meteor's arrival, suggesting that cataclysm and genesis coexist within a paradigm that blurs the line between pre- and post-apocalyptic.

In his analysis of narratives from antiquity to the twentieth century, Frank Kermode posits that depictions of apocalypse are rooted in the individual's desire to imbue their lives with meaning. Imagining the conclusion of history provides a sense of orientation otherwise absent from the human experience (Kermode 7). In more recent texts, apocalypticism increasingly incorporates social critique (Kermode 27). Indeed, *Death and the Meteor* reflects the "irreducibly intermediary preoccupations" of a globalized world decades into the Great Acceleration (Kermode 7).[9] The novel's exaggerated portrayal of environmental decay and interethnic conflict offers an explanation for the impending apocalypse, engaging the critical impulse of dystopian fiction to warn the reader that their society may reach a similar denouement.

Such questions of societal collapse are lamentably quotidian for contemporary Amerindians. In her analysis of science fiction by Native American authors, Grace Dillon underscores the idea that apocalypse has already occurred for these communities. As a result, they experience a continual metaphysical imbalance (9). In Dillon's analysis, native authors respond by imagining optimistic futures with a restored cosmic equilibrium (8–9). Such works embody what literary critic Gerald Vizenor calls survivance, an Indigenous "narrative resistance to absence, literary tragedy, nihility, and victimry. Native survivance is an active sense of presence over historical absence … [and] a continuance of stories" (1). The dystopian Brazil of *Death and the Meteor* is an unbalanced world of normalized exploitation and untethered extractivism. Nonetheless, Terron outlines a possible cosmological restoration that may, ironically, depend on global cataclysm.

The novel does not, however, represent a typical case of survivance. Terron, who is not Indigenous, largely refrains from directly representing Kaajapukugi perspectives and instead presents the tribe through the recollections and observations of Boaventura and the narrator. The resultant absence of Indigenous voices might initially appear to be an acritical reflection of the marginalization of native knowledge that the text purports to critique. However, the blatant untrustworthiness of the supposed expert Boaventura underpins a critique of Indigenous representation that applies both intra- and extra-textually. The discreditation of the Brazilian character and his research encourages the reader to remain skeptical of depiction of the Kaajapukugi within the novel. The same dynamic extends to Terron's own status as a non-Indigenous author integrating a recognizable if fictionalized Amerindian culture into his work. By filtering representations of the Kaajapukugi through the narration of flawed outsiders, Terron implies the influence of his own perspective on his Indigenous characters. Moreover, the lack of native voices becomes increasingly obvious as the narrator discovers the hermeneutic capacities of the Kaajapukugi cosmology. In this way, Terron highlights the omission of Indigenous perspectives without infringing on native storytelling, tacitly suggesting that a curious reader might benefit from seeking out narratives of survivance by Amerindian artists.

Although *Death and the Meteor* is not strictly post-apocalyptic, the novel negotiates the theme of hope common among works of this genre. As Virginia Heffernan describes, narratives set after apocalypse generally consider the possibility of ongoing hopefulness even if some such works ultimately discard the possibility of collective renaissance (5–6). Texts broaching questions of indigeneity are no exception. Indeed, native authors' post-apocalyptic condition underpins their search for renewed metaphysical equilibrium (Dillon 8–9). Even as he avoids directly representing native perspectives, Terron demonstrates his concordance with Danowski and Viveiros de Castro's argument that "Indigenous people have something to teach us [non-Indigenous people] when it comes to apocalypses, losses of world, demographic catastrophes, and ends of History" (104). As climate change driven by capitalism makes the future uncertain for human, animal, and vegetal life on Earth, Indigenous responses to extended persecution and pandemic appear especially instructive.

An Uncertain Eschaton: Apocalypse and Cosmogony

Considered in isolation, *Death and the Meteor*'s conclusion appears unambiguously hopeless. The penultimate page finds the narrator and a fantastically resurrected Kaajapukugi, earlier revealed to be Boaventura's son, "looking down at the Earth dividing itself in flames, its tectonic plates decomposing like an immense puzzle whose pieces eventually break away into outer space, a puzzle that will never again be reassembled" (115). From a rationalist perspective, the meteoric impact represents an irreparable cataclysm. The conclusion would thus represent what Greg Garrard calls a "blank apocalypse: an *eschaton* without a utopia to follow" (93). In fact, Terron foregrounds this prospect in the novel's opening sentence: "Today I see what happened as the irrevocable epilogue of colonial psychosis in the Americas, which I wish was just another lie dictated by the winners and not the truth whimpered about another defeat, this time undoubtedly permanent" (11). The narrator immediately defines his story as a conclusive chronicle of loss and alludes to the post-apocalyptic character of contemporary indigeneity. Even so, Terron gradually counterbalances this sense of despondency with an optimistic reading of the Kaajapukugi metaphysics of eternal return.

Before the prospect of a global renaissance becomes clear, however, Terron presents a more pessimistic vision of Kaajapukugi culture. The tribe resulted from the fusion of two peoples who saw their collective bodies as a lizard and a cat, respectively. The combined traits of the two clans offered a path to survival from the ravages of disease and violence but created a strange cultural hybridity that disavows social hierarches, remains largely silent, and eschews defining certain individuals as shamans (22–23). Both muteness and a lack of spiritual leaders represent apparent markers of a culture in decline. Incommunicability is widespread in Terron's near future, yet he specifies that the tribe's lack of spoken language encourages isolation and melancholy (23). The lack of shamans reflects the egalitarianism of Kaajapukugi society but marks an important distinction from real Amazonian cultures. In his testimony to anthropologist Bruce Albert, Yanomami shaman Davi Kopenawa describes himself as a healer whose work encompasses engagement with the intertwined phenomena of physical maladies and metaphysical decay (114). Without the intervention of shamans, unchecked spirits of disease will make the sky fall (Kopenawa and Albert 132–3). Given the importance of shamanism in comparable cultures, the absence of these figures among the Kaajapukugi suggests a possible reason for the novel's global eschaton.

Terron's description of Kaajapukugi religious practice furthers the text's association between spirituality and apocalypse. The foundation of the group's religion is consuming the hallucinogen *tinsáanhán* and experiencing an ecstatic state of communion with tribal ancestors (57–58). Foreshadowing the Kaajapukugi's fate, however, the beetles that produce *tinsáanhán* are already extinct. In an early passage, Boaventura recounts the grave consequences of this extermination: "Given the disappearance of the *tinsáanhán*, their superior world was also swallowed up, and with it their gods, their rituals, and even the three Skies" (24). The Mazateca shaman who facilitates the relocation to Mexico echoes this pessimistic view:

> Among the shaman's worries, what troubled him most was the Kaajapukugi's sacred remedy [*tinsáanhán*]... It would require a wait that was not available to that people condemned to the physical world without the orientation of their god, exiled by environmental destruction … To a priest like El Negro, it was the greatest catastrophe in a horizon entirely punctuated by catastrophes. (26)

Unable to use their fictional hallucinogen to maintain or improve metaphysical equilibrium, the Kaajapukugi are spiritual orphans poised on the edge of blank apocalypse. Facing certain extinction and cut off from their culture's afterlife, the tribe may choose self-annihilation to ensure the end of a cruel world.

The Kaajapukugi might be justified in seeking revenge, but later revelations about tribal cosmology suggest that renewal is their true priority. In the third chapter, the Brazilian anthropologist relays the last female Kaajapukugi's description of the group's cosmogony:

> In the beginning, there was a great explosion in Di-yǐ-wài, the First Sky, and Di-èr-wài, the Second Sky where we live now, and this impact permitted Xikú-feixiguiuán, the Lost Plot, to come from Di-sân-wài, the Third Sky, inside of Tinsáanhán, she said, the Great Beetle, from whence emerged a black cloud of fifty smaller beetles, the Pilots, who defecated on Xéngjié-de-xuìmián-dao, the Island of Sacred Sleep. Upon eating the feces of the fifty beetles, the Lost Pilot also defecated, and from their stomach emerged the ancestors of the Kaajapukugi, she said, and from them emerged the Kaajapukugi currently on Xijiè. Upon breathing in the innards of Tinsáanhán, we momentarily visit the Third Sky, where our ancestors live in eternal harmony with the Lost Pilot, she said, and this encounter is what

> teaches us to stay alive … and the Origin will always repeat itself, she said, because the number of things that make up the world have a limit, and for this number to be reached, Xijiè, the World, must be repeated. (80–81)

This detailed passage describes a metaphysics wherein eschaton, genesis, and history all recur. As Danowski and Viveiros de Castro explain, cyclical cosmologies exist among real Indigenous peoples in the Americas and beyond (74–75). Many Guarani-speaking cultures of central South America, for example, believe in occasional series of apocalypse and rebirth (Danowski and Viveiros de Castro 76–77). However, the fictional group's stoicism in the face of spiritual isolation foreshadows the extreme nature of their cosmology.

The fictional group's apparent passivity contrasts markedly with the outlooks of their real counterparts. The Yanomami, for example, use shamanism to prevent the end of the current metaphysical cycle. Although this Amazonian group believes the sky already fell in the mythical past and paved the way for the current world, disconnect from the spiritual realm portends irreversible catastrophe:

> If the xawarari [spiritual entities associated with epidemic disease] continue to invade our land, the shamans will all die and no one will be able to stop the forest from turning to chaos anymore … Then the ground will split open, and all the trees will collapse on top of each other. (Kopenawa and Albert 405)

The negotiation between shamans, ancestors, and nonhuman spirits is a constant struggle to ensure survival by ameliorating the metaphysical imbalance described by Dillon.[10] Indeed, Kopenawa argues that, with proper spiritual guidance, humankind can yet overcome the threat of pandemic and deforestation that defines the present moment (280). A collective suicide like that of the Kaajapukugi would represent an irrevocable capitulation from this ethics of survivance. For Terron's fictional tribe, on the other hand, the apocalypse is inevitable. Whether the group summons the meteor or merely commits suicide shortly before its pre-established arrival, the Kaajapukugi display no interest in prolonging the current metaphysical cycle.

Good Intentions and Great Evil

Terron's juxtaposition of a straightforward vision of global eschaton and the Kaajapukugi's cyclical metaphysics underpins an ambiguous portrait of apocalypse reflecting both resignation and hopefulness. Analysis of the mythological figure *Hen-zaogao*, the Great Evil, elucidates this productive tension between blank apocalypse and potential renewal. A catalyst for eschaton whose role is never fully elucidated, *Hen-zaogao* initially appears to be a metaphor for colonialism before becoming increasingly associated with Boaventura. In turn, the Brazilian's shifting relationship with Kaajapukugi cosmology encourages the reader to adopt an active posture and consider whether the tribal metaphysics sufficiently explain the events of the narrative.

The final female Kaajapukugi, already deceased, appears to the narrator in a dream during the novel's first chapter.[11] This oneiric figure speaks in her seldom-used native tongue, which the narrator translates to Portuguese as "*o homem branco é o Grande Mal*" (30, italics mine). This phrasing lends itself equally to two interpretations: White people are the Great Evil or a single White man fulfills this role. At first, the former seems more likely. Despite their isolation in central Amazonia, the Kaajapukugi have suffered over a century of infection and persecution by non-Indigenous Brazilians: "The genocide of the Kaajapukugi had been set off in the end of the nineteenth century … which obliged them to continue … in permanent retreat from the fatal persecution of outbreaks of measles and flu brought by the invaders" (19). The ravages of the Capitalocene compound the tribe's misery by making the Amazon uninhabitable. Ultimately, the Kaajapukugi social crisis is attributable to a wide range of non-Indigenous actors. The final turn toward extinction, however, is largely due to Boaventura.

The third time Boaventura surreptitiously consumes *tinsáanhán* during his stay with the Kaajapukugi, he describes feeling observed by something invisible and threatening (73). After recovering, he suddenly kidnaps the young woman. During the initial abduction, she miscarries in a grim passage:

> The Indian [*sic*] remained unconscious that whole day, and her fever spiked the next morning. Upon liberating her from the rest [of the placenta], throwing the bloody slime in the river's current, I couldn't help but think of it as being the substance of a future that would never come to pass. (75)

Having precluded any long-term future for his former hosts, Boaventura keeps the young woman captive in a remote outpost where she eventually contracts measles and dies. Contrary to Boaventura's original intentions, he stages yet another chapter of the brutal, centuries-long epidemic that has devastated tribes like the Kaajapukugi.

Before she succumbs to illness, the young woman haltingly relates the Kaajapukugi mythology to Boaventura in the Yepá-mahsã language. After the previously cited description of a recurrent universe, she concludes by addressing her captor directly: "and you will go upstream to Kaajapukugi [territory] again, and again and once more … and you will remain forever trapped on this river of destruction and rebirth, Hen-zaogao, Great Evil" (81). The linguistic divide between kidnapper and victim and the female character's severe distress make this testimony unreliable. Still, the Brazilian accepts that he is *Hen-zaogao*. Perhaps this belief assuages his guilt or fulfills what Kermode identifies as the common desire to situate oneself at the apocalypse. Still, Boaventura's decision to fulfill his former captive's prediction and deliver his newborn son to tribal territory suggests he has quickly internalized his malevolent role within the Kaajapukugi cosmology.

The anthropologist opts to leave his infant son on the Island of Sacred Sleep, where the Kaajapukugi cultivate and consume *tinsáanhán*. There, he discovers the entombed corpse of a taikonaut and, inadvertently, destroys much of the island. In the end, it is not climate change but rather Boaventura who destroys the sacred hallucinogen. The Brazilian's brief experience in the Amazon represents a microcosm of centuries of invasion and infection. However, this destruction only fortifies his faith in the Kaajapukugi cosmology. When the tribe's surviving men unexpectedly request his help relocating to Mexico decades later, Boaventura agrees. Aware that the Kaajapukugi believe him to be the Great Evil, the anthropologist must know that no good can come from this renewed contact. Still, he assents, suggesting his faith that an inevitable apocalypse is imminent.

A Rupture in Time's Flat Circle? Kaajapukugi Eternal Recurrence

In the opening pages of the final chapter, "Cosmogony," the narrator analyzes Kaajapukugi temporality. Recalling the sole article Boaventura published about the tribe, the Mexican anthropologist describes how "The Kaajapukugi saw themselves as one and as everyone. The father, the son and the grandson were a single Kaajapukugi, simultaneous and eternal in the passage of time" (99). At last, Terron explicates a major implication of the young woman's description: the Kaajapukugi believe in eternal recurrence. In his classic study of this subject, Mircea Eliade describes this concept, also known as eternal return, as "the cyclical recurrence of what has been before" that includes "the periodic resumption, by all beings, of their former lives" (88; 120).[12] For the Kaajapukugi, time is not linear but rather a closed cycle whose events occur ad infinitum. Befitting the colloquial description of eternal recurrence as a temporal "flat circle," eschaton leads directly into genesis. The meteor marks both the end and beginning of each historical cycle.

Belief in eternal return offers a convincing answer to the mystery of why the Kaajapukugi would relocate to Mexico only to commit suicide. Still, the revelation that Boaventura's son is among the last 50 Kaajapukugi adds an intriguing metaphysical wrinkle. The tribe's cosmogony includes a group of 50 beetles arriving at the dawn of time, yet the unnamed son is the only Kaajapukugi resuscitated before the novel ends. This discrepancy may be a detail elided through Boaventura's retelling, or it could allude to a more profound change. If intergenerational relationships are indeed simultaneous rather than sequential, both the Brazilian anthropologist and his son may embody the Great Evil. If he is indeed an avatar of *Hen-zaogao*, the younger man's rebirth casts doubt on the world's impending genesis.

Although he does not discredit the possibility of permanent eschaton, Terron provides considerable evidence supporting a metaphysics of eternal return. Foremost is the spacesuit entombed under the Island of Sacred Sleep that seemingly confirms the taikonauts' presence in prehistory.[13] This mummy corresponds closely with the mythological figure of the Lost Pilot, while the Tiantáng I shuttle seemingly represents the Great Beetle. Moreover, the narrator notes the uncanny resemblance between the female taikonaut and the tribe's last woman, raising the possibility that they are ancestor and descendent or even the same personage displaced in time. Lost in space at the time of the cataclysm, the taikonauts return to Earth at the beginning of each metaphysical cycle. Instead of completing their original mission to colonize Mars, these Lost Pilots repopulate the Earth by giving birth to a new cycle of Kaajapukugi.

Immediately after finishing Boaventura's recorded testimony, the narrator delves further into the ramifications of tribal temporality. In so doing, he asserts newfound belief that

> time could be manipulated, it wasn't irreversible or ironclad like some claimed, and by isolating themselves in the subjectivity of their own temporality, the Kaajapukugi subverted historical and objective time … [and] allowed the astronauts of Tiantáng I to reproduce themselves not in space, but in time, populating the Kaajapukugi territory. (106–7)

This passage conveys the narrator's disorientation as the foundations of his rationalist worldview crumble. However, he makes a logical leap that seemingly misreads the final revelations of Boaventura's confession. Rather than manipulating time, the Kaajapukugi reappear as the universe recycles in an unvarying loop. The tribe's calm demeanor while facing spiritual alienation and impending extinction signifies acceptance that history is an immutable cycle. Still, the unreliability of Terron's characters and the otherness of the Kaajapukugi, regardless of their true ethnic background, imply that the reader cannot know the full veracity of the tribe's metaphysics. The possibility of blank eschaton persists, yet a universe based on eternal return appears more likely by the novel's conclusion. Such an inalterable conception of history offers little room for optimism, yet Terron maintains a core of hopefulness through his engagement with dystopian critique.

Conclusion: Optimism Despite All Odds

Ravaged by disease and displaced by encroaching capitalism, the Kaajapukugi tribe rests at the edge of extinction. Given this background, the final survivors' suicide may be a justified act of revenge that manifests the apocalypse and extends their suffering to the rest of humankind. However, Terron complicates rather than confirms the hypothesis that the Kaajapukugi's self-sacrifice is a fatalistic response to decades of persecution. The tribe's recurrent, nonlinear cosmology may preclude the concept of vengeance. Boaventura is not a trustworthy figure. The Chinese mission to Mars is not just a distinctive, futuristic detail. By the novel's conclusion, irreparable, blank apocalypse and impending cosmic renaissance each appear possible, with the latter unexpectedly appearing more likely.

Even as Terron offers evidence supporting an imminent genesis, he continues to provoke further questions. The novel's explanation of the Kaajapukugi's eternal recurrence is unreliable, having been coerced by Boaventura rather than freely given. Even if the Brazilian accurately interprets the tribal metaphysics, new uncertainties arise about agency, suffering, and hope. Acceptance of eternal return means that history repeats immutably, casting doubt on the last 50 survivors' ability to influence the novel's apocalypse. These men seemingly accept their impending death, but it remains unknown what role if any they play in engineering the meteor's arrival. Moreover, the final Kaajapukugi must believe that their tribe's subjugation after European contact will eventually recur. Their ancestors will again endure centuries of epidemic, while they will suffer through decades of spiritual abandonment and environmental decay. In time, their oppressors will reappear to commit the same offenses again.

Real Amerindian cosmologies offer potential solutions to this dilemma. For example, the Quichua concept of *pachakuti* embodies a vein of "hope in cosmic renewal through catastrophe" that includes a return to pre-colonial harmony after a period of intensified chaos (Danowski and Viveiros de Castro 76). Another iteration of this concept appears among the Mbyá-Guarani, who have adapted their metaphysics to exclude white people from future cycles of eschaton and recreation (Danowski and Viveiros de Castro 77).[14] Terron does not explicitly deny the possibility of a better future after the meteor. Still, the credence he lends to eternal recurrence means that improvements in future historical cycles likely remain impossible.

The Kaajapukugi commitment to predestined affliction is among the novel's most challenging provocations. At the same time, the text reflects Terron's interest in the related question of how to maintain hope in a decaying world. In an interview with Jonatan Silva, he claims that "The apocalypse, which was just a remote possibility before, has now become a permanent norm. I don't see any chance of another successful genesis."[15] In another interview, he tells André Cardoso and Pedro Sasse, "I perceive a certain optimism in the ending of *Death and the Meteor*, the possibility that everything begins again. However, the impossibility of making everything different and the likelihood of merely reproducing all of humanity's mistakes is a condemnation" (350).[16] The brutal legacies of colonialism—genocide, pandemic illness, environmental destruction—will continue unabated until reaching apocalyptic proportions. Despite this apparently hopeless condition, Terron does not advocate resignation but rather appreciates the remote possibility of a new beginning for humanity.

Death and the Meteor's pessimistic future seems an unlikely inspiration for hopefulness. At first glance, Terron eschews the optimism implicit in many dystopian or post-apocalyptic texts. The final pages end in holocaust. Even if a new society begins imminently, the Kaajapukugi principle of eternal return would dictate history's next cycle merely repeats the same course of invasion, infection, and apocalypse. Perhaps a world permanently consumed by the Kaajapukugi's fiery revenge is the most desirable outcome.

Despite this absence of intradiegetic hope, however, the novel's critical impulse reveals its optimism. Terron's dystopian exaggerations call the reader's attention to easily recognizable social problems. Even as the characters seemingly accept the immutability of these issues, readers may apprehend the ongoing possibility of constructive change. Avoiding the fate of Terron's fictional society remains attainable, but only by ameliorating problems highlighted in the text. The impulse for any action lies beyond the domain of the novel, yet its portrayal of a highly undesirable future highlights the need for concerted action. A new, unified engagement with environmentalism, Indigenous rights, and pandemic prevention might yet allow contemporary society to end vicious cycles rooted in colonialism and manifest rebirth without eschaton.

Notes

1 Krenak's choice of date ignores the Viking settlements established roughly five centuries prior in Canada, instead denoting Christopher Columbus's arrival in the Bahamas as the start of the European settler colonial project in the Americas.
2 This chapter employs Capitalocene rather than Anthropocene, following Jason W. Moore's argument that the latter term unfairly homogenizes humanity and obscures "the naturalized inequalities, alienation, and violence inscribed in modernity's strategic relations of power and production" (173). Capitalocene, on the other hand, acknowledges capital's leading role in environmental destruction and the comparatively minimal contribution to anthropogenic climate change by marginalized groups.
3 Joca Reiners Terron (1968–) is a prolific and critically respected Brazilian author whose texts generally fuse realist, memorialist aesthetics with elements from the genres of crime, fantasy, and science fiction. *Death and the Meteor* is the sixth of his seven novels published to date.
4 All translations from the novel are mine.
5 Recent scientific reports declare the rainforest is nearing a "tipping point" that would see its transformation into a savannah (Carrington). Further, Mike Davis affirms that tropical deforestation is a leading source of emergent viruses that could spark new epidemics and pandemics.
6 Documentarian Vicente Carelli's *Corumbiara: They Shoot Indians Don't They?* (2009) recounts the tragic fate of the Kanoê do Omerê. A local rancher massacred most of the tribe's male members in 1985; many of the surviving women subsequently opted for collective suicide by poisoning (Vilaça).
7 Terron's short story "El último hombre del mundo" fictionalizes the case of the Man in the Hole, so named due to his penchant for excavation. The man's tribal origin remains unknown even after his death in 2022.
8 This chapter approaches Indigenous cosmology through what Nicholas Campion calls "Anthropological cosmologies," an "integral part of human cultural and social systems" encompassing a given group's understanding of the composition of the universe and humanity's place within this order (5).
9 The Great Acceleration describes the steep increase in greenhouse gas output after 1945 (McNeill and Engelke 1–4).
10 The Krenak, from Southeastern Brazil, similarly utilize their connection with ancestors to ritually ensure the sky remains suspended (Krenak, *A vida*).
11 Dreams and other visions are an important form of communication throughout the novel, creating a contrast with the fallible spoken word.
12 Eliade's *Cosmos and History* identifies similar beliefs among traditional societies in the Americas and around the globe. In modernity, Friedrich Nietzsche is the figure most closely associated with the idea of eternal recurrence. See Karl Löwith's monograph for more on Nietzsche.
13 Terron includes a sketch in the text to accentuate this unexpected connection (105–106).
14 For cultures including the Yanomami, white people represent the descendants of a group banished in the beginning of time due to transgressions (Danowski and Viveiros de Castro 76).
15 Translation mine.
16 Translation mine.

Works Cited

"Apocalypse, n." *OED Online*, Oxford UP, March 2023, www.oed.com/view/Entry/9229. Accessed 5 May 2023.

Brum, Eliane. "Opinion: The Death of the Last Juma Elder in the Amazon." *The New York Times*, 2 April 2020, https://www.nytimes.com/2021/04/02/opinion/international-world/juma-people-brazil-amazon.html.

Campion, Nicholas. "The Importance of Cosmology in Culture: Contexts and Consequences." *Trends in Modern Cosmology*, edited by Abraao Jesse Capistrano de Souza. E-Book, InTech, 2017.

Cardoso, André and Pedro Sasse. "Entrevista com Joca Reiners Terron." *Revista Abusões*, vol. 12, no. 12, 2020, pp. 344–52.

Carrington, Damian. "Climate Crisis: Amazon Rainforest Tipping Point Is Looming, Data Shows." *The Guardian*, 7 March 2022, https://www.theguardian.com/environment/2022/mar/07/climate-crisis-amazon-rainforest-tipping-point.

Danowski, Déborah and Eduardo Viveiros de Castro. *The Ends of the World*. Translated by Rodrigo Nunes, Polity, 2017.

Davis, Mike. *The Monster Enters: COVID-19, Avian Flu and the Plagues of Capitalism*. E-book, OR Books, 2020.

Dillon, Grace L. "Introduction." *Walking the Clouds: An Anthology of Native Science Fiction*, edited by Grace L. Dillon, U of Arizona P, 2012.

Eliade, Mircea. *Cosmos and History: The Myth of the Eternal Return*. Translated by Willard R. Trask, Harper Torchbooks, 1959.

Garrard, Greg. *Ecocriticism*. Routledge, 2004.

Heffernan, Teresa. *Post-apocalyptic Culture: Modernism, Postmodernism, and the Twentieth-Century Novel*. U of Toronto P, 2008.

"Joca Reiners Terron: O presente brasileiro é distôpico por excelência." *Associação Oceanos*, 2018, https://associacaooceanos.pt/joca-reiners-terron-o-presente-brasileiro-e-distopico-por-excelencia/.

Kermode, Frank. *The Sense of an Ending: Studies in the Theory of Fiction*. 2nd ed. Oxford UP, 2000.

Kopenawa, Davi, and Bruce Albert. *The Falling Sky: Words of a Yanomami Shaman*. Harvard UP, 2013.

Krenak, Ailton. *A vida não é útil*. E-book, Companhia das Letras, 2020.

———. *Ideas to Postpone the End of the World*. Translated by Anthony Doyle, House of Anansi Press, 2020.

Löwith, Karl. *Nietzche's Philosophy of the Eternal Recurrence of the Same*. Translated by J. Harvey Lomax, U of California P, 1997.

McNeill, J. R. and Peter Engelke. *The Great Acceleration: An Environmental History of the Anthropocene since 1945*. Belknap Press of Harvard UP, 2014.

Moore, Jason W. *Capitalism in the Web of Life: Ecology and the Accumulation of Capital*. Verso, 2015.

Silva, Jonatan. "Entrevista: Joca Reiners Terron." *Cândido*, 28 September 2020, https://www.bpp.pr.gov.br/Candido/Noticia/Entrevista-Joca-Reiners-Terron.

Terron, Joca Reiners. *A morte e o meteoro*. Todavia, 2019.

———. "El último hombre del mundo." Translated by Paula Abramo, *Abya Yala*, April 2019, pp. 58–61.

Vilaça, Aparecida. *Morte na floresta*. E-book, Todavia, 2020.

Vizenor, Gerald. *Native Liberty: Natural Reason and Cultural Survivance*. U of Nebraska P, 2009.

6 Corruption and Cleansing

An Eco-Feminist Approach to the Nature/Culture Dichotomy in Naomi Novik's *Uprooted*

Sara González Bernárdez

Introduction

The predominantly Western dichotomy between humanity, or culture, and nature has been argued by many theorists to model after the subject/other dynamic, by which the pole marked as "other" is immediately inscribed as inferior within the implicit hierarchy (see Braidotti, "Four Theses on Posthuman Feminism" 23–24). The long-standing and cross-cultural (though markedly Western) association of woman with nature is condensed in the words of Ursula Le Guin:

> Civilized Man says: I am Self, I am Master, all the rest is Other—outside, below, underneath, subservient. I own, I use, I explore, I exploit, I control. What I do is what matters. What I want is what matter is for. I am that I am, and the rest is women and the wilderness, to be used as I see fit. ("Woman/Wilderness" 161)

This depiction of woman and the wilderness in opposition and subservience to man shows how both nature and woman's experiences have been constructed as foreign, as something "that is utterly other—that is, in fact, to Man, unnatural" ("Woman/Wilderness" 163). In this dynamic, male experiences and activities are

> mediated by a language of domination that ideologically reinforces masculine identity as powerful, aggressive, and separate over and above nature. The language that typifies a woman's experience, in contrast, situates her along with nature itself. She is seen, and accordingly sees herself, as somehow part of it. (Salleh 208–9)

As such, notions connoted as masculine, such as reason, civilization, or culture, are understood as superior to nature and thus to other traits seen as lesser, which are therefore assigned to women: "the emotions, the body, the passions, animality, the primitive or uncivilised … matter, physicality, and sense experience" (Plumwood 19). This chapter focuses on how Naomi Novik's 2015 novel, *Uprooted*, stages this relationship between nature and culture by constructing a secondary world[1] that "disconnect[s] the reader from the culturally and socially constructed

DOI: 10.4324/9781003345770-9

systems of thought that are grounded in the man/nature binary" (Baratta 3). In other words, the chapter argues that Novik represents this oppositional nature/culture dichotomy through the two main characters of the novel, Agnieszka and Sarkan, in order to then deconstruct and reframe said binary from a dynamic of domination to a mutually impactful relationship. As such, though nature is a complex notion encompassing literal and metaphorical dimensions, it is for the purposes of this chapter directly related to the physical world, specifically the landscape around the characters. Culture, on the other hand, is associated directly with human beings and the civilized societies they produce and inhabit. The argument is that Novik's use of fantasy elements serves to introduce and then subvert the stereotypical associations, which this dichotomy carries, as outlined above. In so doing, the novel is argued to convey "a critical ecological feminism in which women consciously position themselves *with* nature" (Plumwood 21), a revolutionary act that deconstructs dominant power dynamics.

In this novel, the fictional country of Polnya is accosted by the progressive expansion of a "corrupted" magical forest known only as the Wood. Though the forest is itself lush and green, it spreads a toxic sickness through its pollen, fruits, plants, and even the waters of its river, which is why it is constantly described in terms of poison or "corruption" (Novik 71). The symptoms of the sickness vary, but generally and eerily resemble those found in zombie dystopias. Crops touched by the Wood grow gnarled and poisonous, and its animal and human victims become aggressive and cruel, driven only by instinct: "It was a slow and dreadful death … The Wood consuming you like rot eating away at a fallen tree, hollowing you out from the inside, leaving only a monstrous thing full of poison, which cared for nothing but to spread that poison onward" (Novik 72).

As such, the Wood not only sickens their bodies but also takes away their sense of self entirely. Again, much like in zombie dystopias, survivors face the moral quandary of attempting to cleanse the victims—risking infection themselves in a likely fruitless task—or simply killing the infected directly. The most common solution is to give the victims quick and merciful deaths by "fire or beheading", which are the only ways to kill them (Novik 68). The Wood could thus be characterized as Stefan Ekman's so-called "evil landscape", the place where the archetypal evil antagonist of the fantasy chooses to dwell, which is often "ugly and unpleasant" (Ekman 196). Its living conditions are inhospitable, "too hot, too cold, or too poisonous for normal life to thrive" (195). However, it is usually not the land itself that is evil, since that would imply "a volition that the land does not generally have"; rather, it is generally portrayed "as a victim of its ruler's evil" (195). This makes Novik's novel a notable exception to this tradition, as it seemingly positions the Wood itself as an antagonist—not a victim, but a perpetrator. Yet this position is complicated by the eventual revelation that there is a sentient, intelligent entity behind the Wood's actions, known only as the Wood-queen. This chapter examines this figure in light of Rosi Braidotti's ethics of affirmation to prove how her presence and background are crucial to shifting the relationship between human and nonhuman by the end of the novel. This ending achieves a state of tentative coexistence between the Wood and the humans of the valley thanks to Agnieszka's application of Braidotti's affirmative ethics. Such a state arguably echoes an eco-utopia, a community that leaves behind the oppositional dynamic between humanity and nature in order to coexist with and grow alongside the natural world.

Culture and Nature: Agnieszka and Sarkan

The novel's main setting is the valley where both Agnieszka and Sarkan live, very close to the Wood, which makes human intervention in the natural environment seem rather minimal throughout the story: the villages are small and their livelihoods rural and modest, which the novel depicts as human life blended closely with the natural world. This is starkly contrasted with the only city in the story, the capital city of Kralia, which—as Ekman argues—provides "a limit or boundary that … is not transgressed or permeated" by the natural world (129). This is shown through Agnieszka's first impression of the city:

> Wide rich fields planted with grain seemed to go on forever in every direction, flat and unbroken, the whole shape of the world gone strange. There were no forests here … [T]he vast sprawl of Kralia, the capital: yellow-walled houses with orange-brown roofs blooming like wildflowers around the banks of the wide shining Vandalus, and in the midst of them Zamek Orla, the red-brick castle of the kings. (Novik 229)

This description conveys Agnieszka's shock upon encountering a landscape where "human—or cultural—*control*" has taken over the natural environment (Ekman 132). The city dominates the picture and nature is reduced to the controlled, inoffensive presence of the fields. This strikes Agnieszka as unfamiliar or even uncanny. The city seems too "large and strange", like "a picture in a book, not something real" (Novik 230). The novel thus presents us with a clear contrast between nature (represented by the countryside, the forests, lands, and most evidently the Wood) and culture, represented by the human beings who carve a space for themselves within it. The conflict between these two positions is staged through the main characters of the novel, Agnieszka and Sarkan. Their personalities and behaviors are shaped to follow those associated with each of the poles, as previously outlined by Le Guin in the introduction, thus evidencing the gendered nature of the binary in their respective responses to the threat of the Wood—yet also showing the true complexity concealed behind the binary's apparent simplicity.

Sarkan acts as lord of the valley and lives in an isolated tower, without doors or a retinue—only one girl of seventeen to cook and clean for him, whom he chooses from the valley's villages every ten years in exchange for his protection from the Wood. He uses his magic to stop its expansion and mitigate the effects of its sickness, while conducting research in the hopes of defeating the Wood for good. As a character, then, Sarkan represents the "Civilized Man," which Le Guin spoke of ("Woman/Wilderness" 161). He spent his youth in Kralia, where the wizards of the king's court are trained and gather for important matters. Courtly society and the academic study of magic have shaped Sarkan's personality into the reclusive, ill-tempered, and stiff magician Agnieszka meets. His character is thus fluent in what Le Guin calls the "father tongue", the language of authority, of control, and "of power" ("Bryn Mawr Commencement Address" 147). This is illustrated most particularly through the way he performs magic: his spells are long, complex, and reliant on thorough research and discipline, which also represents the highly intellectual, male-dominated tradition he adheres to and is a product of. The presence of magic, a quintessential feature of fantasy, is thus used here to convey the character's personality as well as his views, since it represents Sarkan's attempts to impose order over the world around him. Additionally, this attitude and background explain his deep-set prejudice against the people of the valley. His position as liege lord already establishes a deep chasm in terms of social class between him and the peasant folk who depend on his protection. However, Sarkan's dismissive views of the valley people as backward and uneducated also come, partly, as a consequence of his disdain for their attachment to the natural world, which he sees as "folly" (Novik 395). Therefore, Sarkan chooses to keep a deliberate distance from them: "He didn't try to win our love at all. None of us knew him" (Novik 9–10).

As Sarkan's counterpart stands Agnieszka, a peasant girl who grew up in the valley and hence in very close contact with nature, as well as with the Wood itself. Sarkan selects her to be taken to the tower due to her innate magical ability, which must be trained. While she bore first-hand witness to the horrors and destruction the Wood causes, the novel emphasizes Agnieszka's appreciation of this peasant way of life, despite the hard work it entails. She is shown to love exploring and foraging in the forest, as well as helping her family with fieldwork. Unlike Sarkan, she experiences the world through what Le Guin calls the "mother tongue", described as "repetitive, … earthbound, housebound. It's … common speech, colloquial, low, ordinary, plebeian, like the work ordinary people do, the lives common people live" ("Bryn Mawr Commencement Address" 149). These are clearly perceived to be positive qualities, as they entail communication with and care for others. As with Sarkan, Agnieszka's closeness to the mother tongue is reflected in her peculiar way of performing magic. While Sarkan's need to be in control is reflected in the exactitude of his spells, where each step is carefully performed to achieve predictable outcomes, Agnieszka's spells work by instinct and consist of simple words and repetitions. When Sarkan is infected by the Wood after it attacks Agnieszka's village, Agnieszka manages to heal him—a feat she should not have been able to manage at her level of skill—by singing a plain two-word spell to the tune of "Many Years," a birthday song. Instead of trying to accommodate an artificially imposed method, she resorts to what feels familiar and soothing, using tunes and gestures that feel like the right thing to employ in each moment. The fact that she succeeds shows that these simple workings, which Sarkan disdains as useless, actually hold power beyond his comprehension. Significantly, Agnieszka first finds these spells in "a journal written in a tiny crabbed hand … full of abbreviations" and vague instructions (Novik 83). The journal is later revealed to have belonged to the mythological witch Jaga, a female practitioner working in the margins of the more dominant magical practices Sarkan adheres to. Jaga's work therefore creates a counter-discourse, using the apparent simplicity of the mother tongue to question the dominant male tradition. Agnieszka finds Sarkan's spells constricting and stifling because she cannot express herself in nor adapt to the father tongue; her magic is best channeled through the "repetitive … ordinary, plebeian" language of the other, of nature (Le Guin, "Bryn Mawr Commencement Address" 149).

Agnieszka is thus clearly grounded in the materiality of her body and of the world around her: against Sarkan's rational mind, constantly attempting to subjugate and control the world around him, Agnieszka represents the emotional body, which seeks to communicate, to connect. Sarkan consciously detaches himself, refusing to engage with the valley and its people—not only because of his aforementioned disdain for them, but because he is afraid of "being bound to the valley" (Novik 356). By being born and raised within the valley, its people develop a bond with it which Sarkan can use to channel his magic. Thus, rather than use his own power "to siphon away some of the Wood's strength", Sarkan kidnaps the girls and uses their bond as a means to weaken the Wood's ability to spread its corruption (Novik 355). This keeps him untethered, safe from the Wood's sickness and from the wilderness he so clearly mistrusts, yet also severs him from human connection. When a growing intimacy begins to develop between him and Agnieszka, Sarkan is scared of it and repeatedly pushes Agnieszka away, so that the romantic relationship cannot develop properly. Instead, he seeks the safety of isolation and the comfort of control, craving "objectivity because to be subjective is to be embodied, to be a body, vulnerable, violable. Men especially aren't used to that," and hence their insistence on separating themselves from nature in favor of the presumed rationality and objectivity of culture (Le Guin, "Bryn Mawr Commencement Address" 151). Agnieszka, on the other hand, is not afraid of the vulnerability of embodiment. Instead of seeking detachment, she willingly attaches herself to a close-knit community of family, friends, and neighbors, proving Le Guin's affirmation that the mother tongue "is conversation, … language not as mere communication but as relation" ("Bryn Mawr Commencement Address" 149). Furthermore, Agnieszka was always particularly attuned to nature: "I'd always been able to glean more nuts and mushrooms and berries than anyone …; I could find late herbs in autumn and early plums in spring" (Novik 55), as well as untainted blackberries during an outbreak of the Wood's sickness. This strong sense of connection does partly come from the aforementioned magical bond that ties her to the valley, but she actively chooses to love it, because "[it] was home" (Novik 281). While her general clumsiness and lack of refinement deeply irritate Sarkan, as her general unkemptness is an affront to the artificial perfection he seeks, it is this imperfect simplicity that Novik employs to demonstrate Agnieszka's intuitive ability to understand and communicate with the natural, the material. This ability is what fuels her magical skill and helps her work beyond the limitations of Sarkan's rigid magic—for example, by cleansing her friend Kasia of the Wood's corruption, which was thought impossible to cure.

Agnieszka and Sarkan's characterizations, initially presented in this oppositional manner, are reminiscent of reductive and essentialist understandings, especially of woman's relationship with nature. As Plumwood points out, there is "a romantic conception … that women have special powers and capacities of nurturance, empathy and 'closeness to nature', which are unsharable by men" (8). Since Agnieszka's unbridled love of nature and her attachment to the valley and its people are clearly offset against Sarkan's inability to see beyond his preconceptions, the novel can be argued to reproduce these stereotypes in potentially harmful ways. Presenting Agnieszka's character as attuned to the natural world acknowledges the positive impact which nurture and understanding can have, and thus "recognises strengths in women's way of being" (Plumwood 9). However, it also "fails to recognise the dynamic of power" between nature and culture (9), which maintains its hold over both protagonists. It is not possible to consider how Novik subverts the binary without also acknowledging how these clear associations of women with nature and nurture, and of men with violence and rationality, could reinforce it.

Embodied Nature: The Wood-Queen

In spite of these potential problems, the novel does not vindicate these connections uncritically. Despite their oppositional characteristics, both Agnieszka and Sarkan are more complex than these characterizations initially suggest, as shown by the fact that they share the same goal of defeating the Wood for much of the novel. Sarkan uses his power to support the efforts of the king's army and the valley people to burn the edges of the Wood and purge its victims—curing the milder cases when possible and burning them if not—to avoid the infection's spread. His efforts in studying the Wood are not meant to understand it, only to learn how to best destroy it. Agnieszka, despite her aforementioned closeness to nature, also aids the fight against the Wood in what ways she can. Since she perceives the Wood as dangerous, something that must be contained, she understands the necessity of burning and destroying it, even if she does not like it. Importantly, she does not initially perceive the Wood as driven by evil, but simply as a natural force: "I'd always hated the Wood, of course, but distantly. It had been a hailstorm before harvest … more horrible … but still just acting according to its nature" (Novik 181). However, as the story progresses, it becomes increasingly obvious that this is not what the Wood is. Rather, it is a rational entity—one that spreads its sickness deliberately and strategically, "reaching out the full force of its malice" to hurt human beings (Novik 181). In light of this revelation, Agnieszka begins to hate the Wood in earnest, which complicates her apparent alignment with the natural world. For the sake of saving her village and her people, "I did want to ride into the Wood with axe and fire. I wanted to … call up armies on either side, and raze it to the ground" (Novik 182). Yet, as shown by the grim development of the plot, this violent response does not lead to a resolution, only to what Plumwood calls the master model—the systematic "exclusion and domination of the sphere of nature by a white, largely male elite" (Plumwood 23).

This elite, which Sarkan both serves and represents, is embodied in Prince Marek and his armies. Marek's discourse of hatred positions the Wood as the other—a monstrous aberration that can only corrupt and harm, and which must therefore be controlled, even if subduing it entails extreme violence and the sacrifice of human lives. This violent approach metaphorically represents man's "ancient, profound, and violent" fear of nature (Le Guin, "Woman/Wilderness" 163). This is especially evidenced after Agnieszka discovers that a spell called *Luthe's Summoning*, when cast by both Sarkan and herself, can drive the Wood's corruption out of a person. As the sick person is cured, one of the Wood's heart-trees—the core sources of its power—is destroyed in the process. Sarkan instantly proposes weaponizing this discovery, focusing on the damage that may be done to the Wood instead of the potential lives that could be saved. Therefore, such violence is driven by fear of the Wood's power rather than by a desire to ensure the people's safety. The immediate turn to violence is not necessarily irrational, given the traumatic effects of the Wood's sickness on both victims and survivors. It is, however, rooted in a visceral response to the wildness and unpredictability of the natural world, which humans fail to understand. Such failure leads to this constant turn to taming and controlling—and, if such attempts fail, to destruction. However, and though the Wood's antagonistic malevolence seemingly justifies this approach, Novik continually writes against such brutality. These acts of violence, selfishness—like Prince Marek's ill-fated attempt to save his mother, Queen Hanna, from the Wood—and self-isolation—like Sarkan's—are depicted as fruitless, and lead nowhere except to further violence as the Wood grows increasingly powerful. This is because such actions deny human beings' "vital interdependence on others" (Braidotti, *Posthuman* 234) and enforce a rationalistic individualism that entraps humans, the valley, and the Wood in an unending cycle of mutual violence. Hence the reason why, despite his power and knowledge, Sarkan remains unable to find the definitive answer to explain the Wood's behavior and stop its unceasing spread.

Agnieszka, on the other hand, remains more ambivalent throughout the conflict and is reluctant to use her power to assist in Marek and Sarkan's violence. She thus perceives the cyclical nature of this conflict just as the novel takes a deeply tragic, brutal turn. As mentioned earlier, the Wood differs from the evil landscape as Ekman describes it by the fact of its sentience. It is not a landscape where evil resides, but seemingly the landscape itself acting with evil intent, deploying complex strategies with the ultimate goal of leading humanity to its destruction. When Queen Hanna is rescued, for instance, her own mind is long gone—the Wood has taken control of her body and actions as though she were a puppet in order to extend its sickness as far as the king's court of Kralia. This causes the death of most of the royal family and leads the distraught Marek into a misguided declaration of war against the neighboring kingdom of Rosya, all of which plays into the Wood's ultimate aim: "He'd put on his crown and march east, and after he'd spent our army slaughtering as many Rosyans as he could, the Wood would devour him, too, and leave the country torn apart" (Novik 325). Realizing that further violence will be useless to stop both the Wood and Marek's army, Agnieszka retreats to Sarkan's tower, where they cast *Luthe's Summoning* in a desperate last stand. Unexpectedly, the spell reveals the source of the Wood's curse through a vision showing an ancient, nonhuman woman, "her skin patterned like alder wood and her hair a deep green nearly black, threaded with red and gold and autumn brown" (Novik 383). Due to her apparent position of power, she is henceforth known as the Wood-queen and revealed to be the entity behind the Wood's deliberate spread of sickness and death. Her hatred of humankind is also explained when, in the novel's climax, Agnieszka and Sarkan enter the Wood. The Wood-queen attempts to trap Agnieszka within one of the heart-trees, which contains the soul of the Wood-queen's sister, Linaya. She shows Agnieszka a vision of the Wood in a very distant past, when it was a healthy, peaceful wilderness, nothing like the cursed forest Agnieszka knows.

Linaya explains that these humanoid, dryad-like creatures, the "Wood-people", had lived alone in the forest for many centuries, so that they "began to forget how to be people" (Novik 411). Their Queen thought to establish an alliance with the human kingdom, so that they may learn from one another: "She thought we could be renewed, and teach them in turn; we could give each other life. But they were afraid" (411–12). Linaya shows Agnieszka how the humans betrayed the Wood-queen, entombing her alive and trying to burn the whole forest down, demonstrating how violence against nature comes as a consequence of irrational fear. Not just a fear of the Wood-people's magic, which they do not understand, but also a fear of leaving humanity behind, as the Wood-people seemed to be doing. Rather than try and fight the humans, however, the Wood-people chose to transform completely into trees: "if we stay, if we fight, we will remember the wrong things", Linaya explains, "and then we would become—" (Novik 414). Though she does not finish the sentence, it is implied that what they feared was to become like the humans themselves—violent and fearful. The fantastic setting thus makes it possible for them to reject humanity entirely as they literally become part of the natural world.

This scene shows humanity deliberately turning away from a mutually enriching relationship with nature in favor of one of direct opposition, of exploitation, which begins the cycle. When the Wood-queen escapes her imprisonment, she finds the grove her people became and attempts to turn into a tree herself to be with them again. Her transformation is abruptly stopped, however, when she hears the sound of woodcutter's axes tearing down a tree. Horrified, the Queen kills one of the woodcutters and forces the other inside the tree they were chopping; the man's life-force revives the dead tree, yet his suffering poisons it. Heedless of what she has done, the Wood-queen continues to kill and kidnap humans to plant more trees. The Wood-people had changed willingly, but those taken by the Wood-queen did not: their minds stayed human as they were swallowed by the trees, and their misery and pain began to corrupt the forest. These violent actions—driven by the Wood-queen's desperate rage and misguided desire to protect her people from harm—render her unable to become a tree, symbolizing her inability to achieve peace after the trauma inflicted on herself and her people. Trapped in this stagnant, miserable state, the Wood-queen eventually began to manipulate and direct the whole forest against humankind to attempt to eradicate its threat. This explains why the Wood grew into the fearsome, poisonous, cursed forest Agnieszka knows, yet it also creates empathy for the story's apparent antagonist. By using the premise of fantasy, which allows the Wood-people's existence, Novik is able to literally humanize nature, giving it a body and mind that experience pain, grief, and anger. Furthermore, it is by this use of the fantastic setting that nature is able to retaliate, to fight back and exact revenge on the human beings who have betrayed and mistreated it.

Ending the Cycle

These events can therefore be read as a re-staging of the human/nature dichotomy through the use of a fantasy setting, wherein the humans' strike against the Wood-people becomes a metaphor for human attempts to control and subdue the perceived otherness of nature. This dynamic can be best understood through

Braidotti's notion of an ethics of affirmation, which posits that "ethical evil is equated with negative affects" (*Posthuman* 153). However, in Braidotti's terms, negative and positive affects are "not to be understood as dialectical opposites", but rather "as transversal, non-human forces that need to be assessed in terms of their impact on subjects and on the world" ("Joy" 221). Unethical acts, such as "a hurt, a shock," or "an act of moral, epistemic or emotional violence" harmfully impact an individual's relational capabilities, destroying "the self's capacity to act as they harm the self's ability to inter-relate to others—both human and non-human—and thus decrease one's relationship to the world" (Braidotti, "Joy" 222). In Novik's novel, humanity rejects the Wood by its first betrayal of the Wood-queen, an act of violence that marks the beginning of the endlessly repeating cycle that both Agnieszka and Sarkan become trapped in. The Wood-people consciously rejected these negative affects by turning themselves into trees, refusing to fight and embracing nature as their resting place. The Wood-queen, on the other hand, did respond with violence: she "had to stop them," for humans would continue to cut the trees down and destroy the forest if she did nothing to prevent it (Novik 419). This choice to take revenge, however, harms her "capacity to relate to others … and thus to grow in and through others" (Braidotti, *Posthuman* 154)—quite literally, in this case. She is prevented from transforming into a tree, as her people did, by the bitterness of her sorrow and hatred: "She'd killed the tower-lord and his soldiers, she'd planted all the fields with new trees, and she'd come here with her hands bloody, to sleep with her people at last. But she hadn't been able to take root" (Novik 419). As her sister Linaya had put it, the Wood-queen remembered only the "wrong things", "how to kill and how to hate", negative affects that rendered her unable to change, just like the humans, and made her forget "how to grow" (419). This characterization is reminiscent of the wild, untamed, irrational parts of nature, which have often been used against women in order to justify their oppression. In the same way, human beings use these attacks to justify their violence against the Wood. This is how the Wood-queen, and the Wood itself, begin to be regarded as a purely evil, antagonistic force that must be eradicated.

Furthermore, as the Wood-queen continued to respond with violence, more humans became afraid and lashed out against the Wood, causing the cycle of negative affects to begin anew. This destructive relationship can only be reframed through the affirmative or joyful ethics which Braidotti argues for, and which is staged in a climactic moment of the novel. After emerging from the heart-tree, Agnieszka is given a choice: she can attempt to kill the Wood-queen, drive axe and magic through the core of her heart-trees, and destroy the Wood for good—or she can attempt to help her transform. Though Agnieszka admits to wanting the Wood-queen "to burn, the way so many of the corrupted had burned, because she'd put her hold on them", she nevertheless realizes that "wanting cruelty felt like another wrong answer in an endless chain" (Novik 403). To strike against the Wood-queen in her weakness would only perpetuate a cyclical, desperate war with no victors; against this, affirmative ethics "offers an alternative structure of dealing with difference, which is not oppositional" but rather relational, "able to tell the difference, but to recognize it positively" (Braidotti, "Joy" 222). Agnieszka thus chooses to reframe her relation to nature by reaching out her hand and helping the Wood-queen complete her transformation, which consequently turns "negative relations and passions into affirmative ones" (222). The spell she uses to aid the metamorphosis is in fact a simpler, smaller version of a spell which Sarkan had taught her to change her clothing into fancy, yet stifling, gowns. This mother-tongue version of the spell, combined with Agnieszka's kind gesture of giving water to the Wood-queen, is what breaks the cycle and finally enables the transformation: "I touched her skin and said softly, very simply, '*Vanalem*'. And she was changing" (Novik 420). Her ability to work beyond the limits of the father tongue is therefore what enables Agnieszka to communicate with the Wood-queen, to understand her position, and reach a mutually beneficial solution. Agnieszka is the only character able to put an end to the cycle precisely because she chooses relationality, to both other human beings and nature itself. Instead of violence and control, she applies the positive affects of compassion and understanding, an "ethics of care" that constructs an alternative encounter with the other (Estévez-Saa and Lorenzo-Modia 141), with the Wood she had previously hated and feared. Significantly, the Wood-queen is only able to finally transform by allowing herself to let go of her past grudge and letting Agnieszka's gesture reach her. The scene emphasizes the importance of mutual connection, of allowing oneself to be vulnerable with and relate to another. It is noticeably thanks to the novel's use of the fantastic that this direct connection with the nonhuman is made possible, as the supernatural figure of the Wood-queen constitutes a personified representation of the Wood, and of nature as a whole, whom Agnieszka can speak to and attempt to understand. This would not be possible in our so-called primary world (see Tolkien 52), so that Novik forces a reassessment of readers' perceptions of the nonhuman by using fantasy to depict nature as having a consciousness, an inner life or human-like existence that may be interacted with and related to.

After these events unfold, Agnieszka moves to a small cabin within the Wood itself in order to dedicate her efforts to its restoration and healing, so that it can become part of the valley she loves as her home. She begins purifying the remaining heart-trees by helping the human souls within them "slide into the long deep dream", or else by "giving them to the fire" (Novik 427). Importantly, this burning is not meant to destroy, like those carried out by Sarkan and the king's men, but to give release and relief: it is "the gentlest way" Agnieszka has found to set the more extreme cases free from suffering (427). This portrays the practical application of Braidotti's ethics of joy, "a pragmatic engagement with the present in order to collectively construct conditions that transform and empower the capacity to act ethically and produce social horizons of hope and sustainable futures" ("Joy" 223). By nurturing and caring for the Wood, Agnieszka heals the negative affects that had cursed it, slowly stopping the spread of its sickness. In doing so, she also sets an example for the community around her: her actions, and their perceivable positive impact, encourage the people of the valley to let go of their previous suspicion and fear in order to enter a symbiotic connection with nature, relating to it "in a productive and mutually reinforcing manner" (Braidotti, *Posthuman* 153). Through this deliberate act of care, then, Agnieszka begins to undo "existing conditions so as to actualize alternatives" that do not involve the perpetuation of an oppositional dynamic of dominance (Braidotti, "Joy" 222). Thus, the reframing of the relationship between humanity and nature from an antagonistic duality to mutual understanding and care is crucial to the novel's ending. When the Wood-queen becomes able to let go of negative affects the conflict is brought to an end, showing that coming to an understanding with the natural world was the key to ending the threat of the Wood. However, Agnieszka goes beyond this into healing the Wood as a whole, and in so doing consciously positions herself "*with*" nature, as Plumwood put it (21). She promotes a connection with nature that abandons the master model dynamic and favors mutual coexistence. Humankind and the natural world are no longer "bound negatively by … the guilt of ancestral communal violence" exercised by humans against the Wood-people, "or the melancholia of unpayable ontological debts"; instead, they become linked "by the compassionate acknowledgement of their interdependence" (Braidotti, *Posthuman* 39). If we accept Fátima Vieira's concept of utopia as "a kind of reaction to an undesirable present and an aspiration to overcome all difficulties by the imagination of possible alternatives" (7), Agnieszka's purpose to cleanse the Wood can be read as an alternative to the primary world's domination and exploitation of nature. The novel's closure thus gestures toward an ecological utopia, representing the slow creation of a community that favors respect and understanding of nature, and learns to grow through and alongside the natural world in order to thrive.

Sarkan, on the other hand, still mistrusts the Wood; he sneers at Agnieszka when she announces her intention to cleanse it, and leaves her behind without another word. She understands that Sarkan is not wrong in his misgivings, as the forest remains dangerous despite the Wood-queen's passing; Agnieszka also knows, however, that he did not leave "for the sake of corruption or the kingdom", but because "his tower was broken, he'd drunk Spindle-water, and he'd held my hand" (Novik 423). In other words, he left because, despite his best efforts, he has established a connection with both the valley and with Agnieszka herself. He remains clearly afraid of this vulnerability, this perceived defect in his rational, controlled relationship to the world, and so he has "run away as quick as he could", to "keep himself locked away for ten years this time, until he withered his own roots" (423). Agnieszka, on the contrary, does not completely isolate herself from the rest of the valley. She instead becomes the prototypical figure of the witch in the woods—much like her mentor Jaga—, and thus a parallel to Sarkan; yet she still comes out of her cabin to help wherever she is needed and visits her family and friends often. Despite her choice to live in close communion with nature, Agnieszka never foregoes her connections to other human beings, representing the balance between nature and culture which she attempts to sustain.

Conclusion

While Agnieszka and Sarkan represent the binary between nature and culture as a relationship of opposites, the evolution of the plot and of their relationship shows that the dynamic is far more complex than the simple binarism can convey. Despite reproducing behaviors associated with the binary, both characters deviate from these monolithic categories. Agnieszka is not inherently nurturing, as she does hate the Wood and desire retribution; and, despite his ingrained beliefs about the Wood and the valley, Sarkan still chooses to help Agnieszka achieve her goals. Such complexity is reflected in the larger-scale conflict between human beings and the Wood. Just as human beings are not only victims of the Wood's sickness, but also its root and perpetrators of the violence it suffered, the forest is not the evil, antagonistic force it appears to be, but a desperate being in need of compassion and closure. When faced with the truth about the Wood, Agnieszka makes a conscious decision to let go of her hatred and grudges against it. Instead, she chooses to remain connected with nature and to try to understand it, a choice that brings about healing rather than further destruction.

Uprooted therefore stages the power dynamic between humans and nature by personifying the elements which form this duality and playing out their conflict within a secondary world. At the end of the novel, paralleling a scene from its beginning, Sarkan returns to the village to find Agnieszka, showing his willingness to not only accept his connection to her but also create more with others, and thus allow himself to grow "in and through others" (Braidotti, *Posthuman* 154). The gap between nature and culture is bridged in this final scene as Agnieszka reaches out once again, this time to take Sarkan's hand. This act represents their willingness to work past the differences that had separated them. Moreover, it metaphorically reinforces the connection between the human and the natural, which Agnieszka has made possible. It is therefore possible to make a reading of *Uprooted* as an eco-feminist novel. While Novik does not make a straightforward argument in this regard, the plot of the novel encourages coexistence with and care for the natural world. Furthermore, Novik's personification of nature through the Wood-queen, and Agnieszka's act of reaching out to help her, conveys what have been some of the major principles of eco-feminism—a recognition of the interdependence between human and nature and of the importance of the impact they have on one another. Thus, the almost eco-utopian turn, which Agnieszka is working toward, shows how the application of affirmative ethics could potentially help shift the oppressive dynamic of nature vs. culture, creating instead a community based on coexistence and mutual care that focuses on healing and restoration.

Note

1 Following J. R. R. Tolkien's terminology (52).

Works Cited

Baratta, Chris, editor. *Environmentalism in the Realm of Science Fiction and Fantasy Literature*. Cambridge Scholars Publishing, 2012.

Braidotti, Rosi. "Four Theses on Posthuman Feminism". *Anthropocene Feminism*, edited by Richard Grusin, University of Minnesota Press, 2017, pp. 21–48.

———. "Joy, Ethics of". *Posthuman Glossary*, edited by Rosi Braidotti and Maria Hlavajova, Bloomsbury, 2018, pp. 221–24.

———. *Posthuman Knowledge*. Polity Press, 2019.

Ekman, Stefan. *Here Be Dragons: Exploring Fantasy Maps and Settings*. Wesleyan University Press, 2013.

Estévez-Saa, Margarita and María Jesús Lorenzo-Modia. "The Ethics and Aesthetics of Eco-caring: Contemporary Debates on Ecofeminism(s)". *Women's Studies*, vol. 47, no. 2, 2018, pp. 123–46.

Le Guin, Ursula. "Bryn Mawr Commencement Address". *Dancing at the Edge of the World: Thoughts on Words, Women, Places*, Grove Press, 1989, pp. 147–60.

———. "Woman/Wilderness". *Dancing at the Edge of the World: Thoughts on Words, Women, Places*, Grove Press, 1989, pp. 161–64.

Novik, Naomi. 2015. *Uprooted*. Pan Books, 2016.

Plumwood, Val. *Feminism and the Mastery of Nature*. Routledge, 1993.

Salleh, Ariel. "The Ecofeminist/Deep Ecology Debate". *Environmental Ethics*, vol. 14, no. 3, Fall 1992, pp. 195–216.

Tolkien, J. R. R. *On Fairy Stories*, edited by Verlyn Flieger and Douglas A. Anderson, Harper Collins, 2014.

Vieira, Fátima. "The Concept of Utopia". *The Cambridge Companion to Utopian Literature*, edited by Gregory Claeys, Cambridge University Press, 2010, pp. 3–27.

7 Through Currents of Contamination

The Failure of Immunizing Insularity in Sophie Mackintosh's *The Water Cure*

Eleonora Rossi

At the start of the millennium, Paul J. Crutzen and Eugene F. Stoermer argued that a new term might be needed to describe the current geological epoch, which is characterized by the heavy impact of human activity on the planet. Thus was the "Anthropocene," the domain of *Anthropos*, man, officially born. Scholars across a wide variety of fields have been grappling with the meaning, impact, and timelines of what has emerged as a haunting image of precarity, decay, and intersecting crises. In this context, it has been argued that the arts and humanities have a particularly central role to play, by raising awareness of ongoing crises and spurring climate action. This piece is situated within this rich interdisciplinary background and is particularly aligned with the emergent and bourgeoning field of the Blue (or Blue-Green) Humanities. It builds on posthuman inquiries into contemporary literature to foreground a specific poetics (and politics) of contamination, which involves water as well as human bodies and lived spaces. More specifically, I use Sophie Mackintosh's *The Water Cure* (2018) as a case study through which to examine instances of dynamic and incontrollable exchange between human and nonhuman worlds. Against the desire for and violent attempts at enclosing and protecting human bodies and lived environments from contamination with external, polluting agents, this novel presents us with instances of unregulated porosity. These provide opportunities for generative, feminist rewritings of planetary existence beyond Anthropos. Though the novel is primarily concerned with bodily, architectural, and geographical boundaries, the considerations of the inevitability of permeation—and therefore of contamination—which are articulated throughout this chapter have broader implications. The barriers that are transgressed during instances of unregulated influx tell a larger story concerning the inadequacy of human utopias of containment. This, as we will see, has critical and political consequences for discourses on the heteropatriarchal ordering of society, both in the novel and beyond, practices of racialized segregation (particularly in relation to the waves of migration that are currently challenging European politics), as well as for discourses on architecture and infrastructure as we live through the rippling effects of climate change.

In a recent publication, American philosopher Wendy Brown explores what she terms a fundamental tension in the Western world between "opening and barricading, fusion and partition, erasure and reinscription" (7–8). Here, Brown warns us

DOI: 10.4324/9781003345770-10

against the dangers and empty promises of enclosure as a strategy of "encampment," which translates into rigid borders, fences, and walls (7). The establishment and preservation of boundaries against a multitude of material and immaterial fiends is central to *The Water Cure*. In this novel, Mackintosh introduces us to a small patriarchal society, initially composed of a single nuclear family of five, that has settled on an unnamed island to escape an epidemic that is allegedly ravaging the mainland. The word "allegedly" is essential here, in that no evidence is procured of the existence of the epidemic other than King's—the family's patriarchal head's—words and experiences. Nonetheless, King, his wife, "Mother," and their three daughters, Grace (the eldest), Lia, and Sky (the youngest), live as if danger hides in every being that washes ashore, as well as in every air particle or water droplet that crosses the island's natural confines.

The threat posed by the epidemic is not evenly distributed: King explains that it is a virus born within (cis) men's bodies, which can make cis women and girls gravely ill. Measures are needed to gain distance from the dangerous mainland and its men, and to strengthen women's bodies and spirits against the infectious incursions. To this end, King takes his family to an abandoned sanatorium on a deserted island, where he develops a series of "cures" (which give the novel its title) that are aimed at cleansing women's bodies from within and toughening them against the possibility of disease—making them once again "healthful and whole" (Mackintosh loc. 122). These practices of pre-emptive enclosure illustrate Adi Ophir's and Ariella Azoulay's argument in "The Monster's Tail" that "[i]t is not the wall that has created the camp, but rather the strategy and reality of encampment which has led to the construction of the wall" (quoted in Brown 7). Yet, King's many walls are fallacious—or even imaginary. On the one hand, his cures, which include torturous practices of self-harm, do not lead to the results desired. Once King's presence wanes, the sisters do not find themselves impermeable to external stimuli, but rather thirsty for emotions and experiences beyond the suffering to which they've grown accustomed. On the other hand, the realization that takes place toward the end of the novel that King's island was, in fact, always a peninsula, invites us to reckon with the impossibility of insulation in the face of unseen patterns and avenues of connection. In both instances, water, which leaks, stagnates, and carries things away from the island while washing new things ashore, is instrumental to the demystification of King's utopia of immunizing insularity.[1]

As mentioned above, King first develops his cures to heal and strengthen his daughters' bodies against the epidemic, which only seems to affect cis women, causing symptoms including fatigue, bleeding gums, hair and weight loss, and even death. King holds the belief that women are significantly more susceptible to the intake of toxins from the outside world, and, for this reason, their bodies must be shielded from contamination and regularly cleansed from within, to ensure the elimination of all poisons. From his point of view, a woman's body "attract[s] harm" by engrossing matters from the outside (Mackintosh loc. 762).[2] Poisons "hook into [women's] hair and organs and blood and become part of [them], the way heavy metals do" (Mackintosh loc. 430). They "sit inside" like "misshapen pearls," calcifying in their "veins and the chambers of [their hearts]" (ibid.).

As the body is unable to filter out noxious substances, its surfaces are framed as a passive "layering of flesh around everything ingested and experienced" (ibid.). His understanding has ancient roots. The somatophobia (fear of the body) that pertains to the female form has historically been tied to the "unrepresentable status of women's body fluids" within prevailing and widely shared ontological models over the centuries (Grosz 192). In line with the medical practices and understandings of their times, Hippocrates (370 BC–460 BC), Aristotle (384 BC–322 BC), Pliny the Elder (23/24 AD–79 AD), and Thomas Aquinas (1225 AD–1274 AD) argued that (biologically) female bodies are more porous than masculine ones and—therefore—"faulty" (Dean-Jones). With some differences, they all held the view that the female body is loosely textured, spongy, and porous, and produces extra fluids that need to be excreted (Dean-Jones 178). Further, in *De secretis mulierum* (*On the Secrets of Women*), a thirteenth-century treatise on human generation influenced by Aristotelian biology, the author constructs a "leaky, contaminating female body whose superfluities threaten the integrity of proximate bodies with wounds, illnesses, and deformity" (Miller iv). All of these attributes—unstable, penetrable, overflowing—were consistent with medieval representations of monstrosity, thus characterizing women as instances of threatening, watery embodiment (Miller iii). Though cultural understandings of the female body have evolved throughout the centuries, nowadays scholars continue to explore cultural ideas that connect bearing a "seeping, leaking bleeding womb" to the monstrous (Andreasen 902). Like the monster, both female and queer bodies still represent dangerous in-betweens, "objects of otherness," that are synonymous with "uncontrollable nature, moisture, pain, and weakness" (Andreasen 902).[3]

King's perspective evokes cultural understandings, which code femininity with dysregulation, a loss of control over both the body and emotions, and consistent cross-contamination between inside and outside, humanity and nature, or animality.[4] This stands in contrast to the framing of male corporeality in the novel as "irreproachable" (Mackintosh loc. 430). Men's survival in Mackintosh's world is depicted as "implicit" and "taken for granted," because the world is "a man's place" (loc. 2309). King upholds a vision of masculine impenetrability as a dominant and superior type of embodiment. This also corresponds to "a bounded materiality that houses a human subject," and which is "contained within…the skin… beginning and ending in the 'I,'" as Astrida Neimanis suggests (1). In King's view, epithelial boundaries are analogous to the confines of the island or the thick walls of the sanatorium: they delimitate a (human) space that is secluded and therefore safe, impenetrable to the ailments that come from outside. The purported porosity of the female form, which allows for unregulated exchanges, threatens the stability of this binary system. Yet, as the recent COVID-19 pandemic amply illustrated, viruses, such as the one that allegedly circulates on the mainland in the novel, do not respond to epithelial boundaries, nor to cultural, political, or social divides. This makes King's dream of protective insulation insubstantial. Further, in arguing that women's bodies are more susceptible to virulent intakes (as well as to the leakage of harmful energies), King does not acknowledge the fact that the poisonous substances are seemingly produced by men's bodies. This makes them dangerously leaky and thus closer to that image of unregulated and fluid monstrosity against which he endeavors to build protective fortifications. This perspective summarizes the sense, developed throughout the novel, that threats were never (only) outside, and the walls always immaterial, imaginary, and bound to disappear.

The tension between outside and inside, the island and the mainland, purity and contamination, order and unruliness, is maintained throughout the novel, hanging over every surface like a cloud. Women undergo the water cure, which sees them strive to cleanse the body from within by means of drinking salt water to produce vomit, and then consuming glass after glass of "pure water…that came from the island's taps" (Mackintosh loc. 209).[5] This repeated practice, however, does not leave their bodies stronger, but rather ravenous and malnourished, their teeth black and falling. As King considers the sea to be too polluted, the sisters only swim in a pool that is "the sea made safe," where salt water is "filtered through unseen pipes and sluices" and "thick rivets of salt are laid down on the tiles immediately bordering the water, guarding against toxins brought in on the wind" (Mackintosh loc. 99). Irrespective of these defenses, toxic air continues to move swiftly across the sea, while the tides fill the shore with rotten, squat catfish and "jellyfish full of poison" (ibid.). Salty droplets from the ocean winds reach and stain the house's windows, leaving unmistakable traces of the water's reach beyond the shore. Despite their attempts at cleansing and control, the sisters' bodies become an environmental pollutant; overcome with the escaping energy of their own, unregulated emotions, Mackintosh describes them as capable of secreting a noxious smog that is described as "clinging around the house, the forest, the beach" (Mackintosh loc. 21–22). Upon King's disappearance, the sisters believe that those energies are responsible for "driving him away" (ibid.).

When King vanishes and his family assumes he is dead at sea, two men (Llew and James) and a child (Gwyl) land on the island's shores and seize control of the territory. From that moment, the illusion of protection constructed around the space of the island implodes. Llew and James counter King's strictures and practices, imposing a different regime onto the island. Yet, theirs isn't a benevolent rule; upon realizing that Mother will continue to resist their presence, they kill her. In addition to this, Llew seduces and (seemingly) impregnates Lia, while also causing her significant emotional distress. Eventually, he reveals to her that it was King, who was still alive and had simply relocated to the mainland, who had instructed them to murder his wife and retrieve his daughters. He had deemed their life on the island a failed experiment which he wished to end. Llew also reveals that the island itself was a fiction, precariously enforced by means of a barbed wire running through the woods that separate it from the mainland.

With the revelation that the island is in fact a peninsula, Mackintosh produces a final critique of King's impossible dream of protective insulation, which he himself even decides to abandon, while commenting on the historical construction of the island as a utopian Eden that provides protection from the ills of the world. As Joseph Meeker points out, "[i]slands are heirs to an ancient mythic legacy that depicts them as the natural home of wonders and feelings of freedom, however illusory these may be in fact" (202). In European history, Eden has always been imagined as insular, and represented as "either landlocked or sea-girt" (Gillis 25). A longstanding literary tradition, which includes not only Thomas More's *Utopia* (1516), Francis Bacon's New *Atlantis* (1626), and Aldous Huxley's Island (1962), but also Charlotte Perkins Gilman's *Herland* (1915), has framed small, unchartered, and isolated islands as socio-political and ecological utopias. According to Jill Franks, islands "exert a unique appeal because they are contained spaces, and therefore, theoretically at least, controllable" (7).[6] Their apparent boundedness and remoteness has encouraged humans and, as Godfrey Baldacchino highlights, "usually men," to unleash their "atavistic desires for power and control" to repopulate the "tabula rasa" of the island world (105).[7] The limited domain of the island has been linked to the human body and the human self; Franks argues that "in literature, the island is a powerful icon of the self...It is easier to imagine oneself an island than a continent" (10). Yet, as both the etymology and the history of islands suggests, these views are built on unsteady premises. The term "island" has a complex history, which ties it to both the Old English *ëgland* ("*ieg*," "a formation on water" and "land"), the French "île" and the Latin "insula" (Klaiber 7). This last term is thought to derive from the Greek *ev aXi ovoa* ("that which is in the sea") (Klaiber 7). Recast as a hybrid formation of earth and water, sea and land, the island is no longer an image of separation or enclosure, but rather one of unseen interconnectedness: it is "in the sea," always entangled, always in flux. In *The Water Cure*, Mackintosh engages with analogous images of apparent containment and concealed connection. By revealing that King's island was never in fact an island, but rather land that is of another land, an earthly extension into a watery world, Mackintosh invites readers to engage with and question the (im)possibility of successful severance and the lingering threat (or potential) that there may always be strings that, though hidden, keep us attached. Similarly, by describing a virus of unclear origin or existence (some have suggested that the purported epidemic represents a materialization of pervasive toxic masculinity, rather than a pathology), she sheds light onto the ubiquity and opaqueness of contagion. In this sense, King's obsession with strengthening both skin and the island's natural boundaries, which act as fallacious shields against threats which cannot easily be apprehended, or boundaries whose failure indicates the impossibility of separation, is fully revealed as the last hurrah for a strategy of enclosure, which Mackintosh suggests has become unavailable as we swim in overlapping and sometimes dangerous streams of inevitable intimacy.

Throughout the novel, Mackintosh thus navigates multiple sites of tension between King's attempts at upholding a vision of salvific, utopian insulation versus unavoidable instances of contamination with unseen agents, including viruses, bacteria, and environmental pollutants, that refuse to be kept out, drawing, every day, "closer to us" (Mackintosh loc. 2476). Despite King's best attempts, all boundaries in *The Water Cure* are presented as "hydrophilic, through and through," always connected with and infiltrated by other bodies and substances (Neimanis 65). This is the awareness that the sisters come to while looking out at the crumbling world around them: they know, deep down, that even being physically removed from the origins of contamination wouldn't save them. This is also Lia's realization when she declares that there is "no staying away" from the toxins in the air and sea, as everything "was already in our blood…border or no border" (Mackintosh 2375, 2699). Their allegedly "safe place" had "been contaminated from the start" (Mackintosh loc. 2375). At the end of the novel, looking out onto the beach, parting the curtains that keep the world away from sight, the sisters realize that "[t]he end is coming […] [w]e feel it, like electricity, like the start of a migraine…everything's becoming ruined" (Mackintosh loc. 1987). Ultimately, this translates into the sisters' decision to walk through the forest and over the barbed wire that marks the entrance to the peninsula, with the intention of re-entering what they imagine to be a community of women on the mainland from which they had been unjustly severed.

On the one hand, the sisters' words gesture toward the unavailability of King's "politics of purity" when the contamination of human bodies and spaces with "carcinogens, neurotoxins, asthmagens and mutagens," among other substances, is inevitable (Hayward 257). This is particularly true if we think with water, as fluid exchanges force us to consider that we swim in a "socially complex soup of pesticides, fuels, process chemicals, solvents, toys, and packages," with which we are always trans-corporeally connected (MacLeod 264). On this point, the realization that the island was always a peninsula, and that the detritus was not (only) coming from the decayed mainland, but also from the open sea, reminds the reader of the profound anthropogenic impact over natural spaces. The already long-polluted ocean no longer represents a uniform and empty expanse as portrayed within the myths of European imperialism. As Isabel Hofmeyr argues, "[y]ou don't see the 'graves' of thousands of drowned slaves marked in the Atlantic. Undersea cables and oil rigs do not appear, nor do islands of plastic particles. You also don't see any trace of the more than one million shipwrecks that litter the ocean floor" (2018). Despite their apparent invisibility, these histories and networks run through and materially shape the ocean, contesting its framing as an empty expanse, capable of erasing humanity's sins. Further to this point, within the Anthropocene and its manifold networks of extraction and pollution, nothing can be "effectively isolated, insulated, instituted, even immunized, as something apart, something that might be considered proper only to itself" (Greg and Short 1). In other words, if there were ever barriers to begin with, we are now living among their ruins.

These considerations have further political implications. Though the exact location of the island is not revealed, we can locate it off the coast of the United Kingdom, possibly near Wales. This is due to both the name choices for Llew and Gwyl, which speak to Mackintosh's own Welsh background, and the fact that the abandoned sanatorium which hosts the family as well as the water cure are described as in line with the practice of "taking the waters," which was popular in the nineteenth and twentieth centuries, particularly among the British social elites (Adams 2). Taking this geopolitical background into account, the tension articulated in the novel between internal purity and external contaminants sheds light onto a number of significant and interconnected issues. Specifically, King's preoccupation with the penetration of the island's shore by polluting agents maps onto longstanding British and Western xenophobic anxieties concerning the arrival of unwanted bodies at borders. In the UK context, this became particularly evident during the 2016

Brexit debates, which were driven by nationalist concerns over waning sovereignty in the face of continued immigration and which led to the departure from the EU.

As Joshua Goldstein and Devid Rapkin stress, insularity and hegemony are often intwined (935). More broadly, the ongoing European Migrant Crisis (2015–), which has seen the Mediterranean and its islands become sites of massive movement of populations as a result of global inequalities, of which climate change is a contributing factor, also speaks of the Western terror of unregulated influxes. Though international migrants represent only a small fraction of the global population and most migration happens within a country's borders, in recent years, a language of crisis has taken root, which infuses migration with a sense of moral panic. This is connected to the purported threat of Western countries being "overwhelmed" by surges of foreign migrants (Hui), who are constructed as threatening to national, cultural, and bodily sovereignty. In this connection, right-wing, nationalist movements across Europe have been threatening to "close the ports" to migrants, dooming them to the whims of the sea and, in several cases, to certain death.[8] Yet, as continuous arrivals (and deaths) demonstrate, these are inadequate measures and paradigms, which call for urgent rethinking. By suggesting that the only real threat in this imagined world is King and his emissaries, Mackintosh contests colonial and racist ideas concerning the intrinsic dangerousness of bodies that are constructed as "pollutants," and whose arrival and permanence purportedly threaten to undo both bodily and national integrity.

The notion of utopia, as seen above, exists in a complicated relation with that of boundaries. On the one hand, we might see the former as a type of enclosure, which is constructed upon the confines that are placed around the fantastical image of an ideal existence. Yet, as a "no place," utopia can also be seen as an opening, and the "site of new beginnings," or dynamic milieu in which "possibilities are created, rehearsed and tested" (McMahon 8). This is aligned with the (re)framing of utopia as utopianism, or the active striving for better worlds that is alive in the praxis of radical social movements and prefigurative communities, among others (see, for example, Moylan 2020). In *The Water Cure*, Mackintosh places these understandings of utopia(nism) in conflict, reflecting on the impossibility and (potentially) violent enforcement of enclosure in a world that resists it and overwhelms it on several counts.[9] As sociologist Zygmunt Bauman puts it, the "[u]topias of yore" stand "condemned" in the interspersed and "globalized world" of today (236–39). Utopia is left "homeless and floating, no longer hoping to strike roots, to 're-embed'… The Utopian model of a 'better future' is out of the question" (Bauman 236–39). In the novel, the sisters' disillusionment with (not to say hatred for) their father's violent, heteropatriarchal dream of insulation and self-sufficiency reframes King's island as a dystopian space. Their emancipatory journey toward the mainland, on the other hand, foregrounds utopia(nism) as a collective endeavor, animated by a shared wish for creating more equitable realities.

For as derelict and dangerous as it may be, the mainland is imbued with a sense of hope, situated in the possibilities of community and collaboration. In the text, this is manifested in the sisters' burning desire to rejoin the ranks of a community of women, whose anger they imbue with the capacity of changing the world:

> It is possible there are no safe places left. It is possible that we can create a new one with our rage and our love, that other women are already out there, doing the work. We are going to meet them. They will recognize us, no longer children, and hold out their arms to us. They will say, What took you so long? (Mackintosh loc. 2697)

While this passage captures some of the key aspects of the novel's utopian message, it also suggests that Mackintosh's novel rests on an exclusionary view of feminism (as highlighted by the fact that no mention is made of women who are not white, not able-bodied, nor whose bodies were not assigned female at birth). At the same time, in its questioning of boundaries, it includes the (unspoken) possibility of broadening the space of collaboration beyond the unsatisfactory and violently enforced confines of cis, white womanhood. As Anna Tsing outlines, using the example of matsutake mushrooms, today's is a rickety world, where stables structures are "thrown into shifting assemblages" (42). In this context, we become aware of our precarity and vulnerability, and learn the lesson of "collaborative survival," that is, the art of staying alive through "livable collaborations" (Tsing 42). "Collaboration," for Tsing, "means working across difference, which leads to contamination. Without collaborations, we all die" (43). The novel's final passages convey an ambivalent message, which (perhaps opportunely) does not resolve the tensions, which had been constructed throughout. First, though the sisters' choice to cross into the mainland points to their intention to continue existing with and within (rather than in denial of) manifold networks of inevitable contamination, their drastic abandonment of the "island" space speaks to a desire, analogous to that of their father, to leave the ruin of their previous home behind and start anew in a supposedly promised land. This impression is strengthened by the repeated mention of "safe spaces" that are left to create, alongside other women who are "doing the work", as seen in the section cited above. At the same time, the worry that there may be "no safe places left" haunts the same passage, problematizing the possibility of hope. The conclusion of the novel with an image of walking ("the three of us, taking step after step") leaves open the possibility of further, unpredictable shifts and evolution. Mackintosh's world is thus always in formation, suspended between forceful enclosures and emancipatory horizons that exist "somewhere past" our current realities, which can be reached provided that one is willing to "walk far enough" (Mackintosh loc. 2210). Mackintosh's ambivalent stance, which oscillates between a posthuman appreciation of more-than-human networks of planetary entanglement and an ecofeminist insistence of the centrality of "women" to the "work" that is needed to change current predicaments, frames this as a novel of the Anthropocene. As Stacy Alaimo discusses, the Anthropocene is a setting that defies and eludes "transcendent, definitive mappings, transparent knowledge systems, or confident epistemologies" (2). Consequently, *The Water Cure* rejects the possibility of certainty and amplifies—rather than resolves—the readers' doubts, in relation to the epidemic, the possibility of salvation, hope. Navigating the uncertain waters of an imagined pandemic, the debatable boundaries of an unreal island, and existent maritime networks of exchange with other bodies and other waters, Mackintosh encourages us to hold abundant space for the emergent complexity, which is born among the ruins of those which were our walls—amid inevitable but possibly generative networks of cross-contamination.[10]

Notes

1 The concept of immunising insularity is developed by Roberto Esposito. See Bird and Short 1.
2 Mackintosh frames Mother in similar terms, by portraying her as having absorbed King's theories: "More than that, she was a woman at our father's side, absorbing and refining his theories" (Loc. 60).
3 The element of uncontrollability is also essential to the novel, where purity is framed both as a physiological state of (in)contamination and as a code of behavior, which defines the rightful state of the nuclear family as one where wife and daughters follow the patriarch's commands.
4 Among other examples that discuss female dysregulation, see also Groneman.
5 In the novel, it is not only the sisters that are subjected to the cures but also women who traveled to island specifically to receive treatment. Whether the women came on their own will or were sent by male family members seems unclear, however, as at least one woman in the novel commits suicide and another one successfully escapes. In the text, no precise reason is provided to explain why, at some point, the women stopped coming.
6 This also evokes the characterization of the island offered by Edmond and Smith, according to whom islands are "the most graspable and the most slippery of subjects" (5).
7 On this point, it is worth noting that More's Utopia—the founding text of utopian literature—is also set on a peninsula. King Utopus converts this into an "island" by having his slaves dig a deep trench to separate it from the mainland.
8 This particular expression, "chiudere i porti," gained prominence among Italian right-wing movements.
9 We may think about environmental pollution, but also of networks of commerce, data and information, and bodies, as well as viruses, as the COVID-19 pandemic demonstrates.
10 On the image of walls, I was inspired by Amy Butt's keynote talk, "Held in Common: Science Fiction and Collective Space," delivered at Un/Building the Future: The Country and the City in the Anthropocene at the University of Warwick in June 2023.

Works Cited

Adams, Jane M. *Healing with Water: English Spas and the Water Cure, 1840–1960*. Manchester, Manchester UP, 2015.

Alaimo, Stacy. *Exposed: Environmental Politics and Pleasures in Posthuman Times*. Minneapolis, University of Minnesota Press, 2016.

Andreasen, Lise Ulrik. "Menstruation Mediated: Monstrous Emergences of Menstruation and Menstruators on YouTube." *The Palgrave Handbook of Critical Menstruation Studies*, edited by Chris Bobel et al., Singapore, Palgrave Macmillan, 2020, pp. 901–914.

Baldacchino, Godfrey. "Islands and Despots." *Commonwealth & Comparative Politics*, vol. 50, no. 1, 2012, pp. 103–20, https://doi.org/10.1080/14662043.2012.642119.

Bauman, Zygmunt. *Liquid Modernity*. Cambridge, Polity Press, 2000.

Bird, Greg, and Jon Short. *Community, Immunity and the Proper: An Introduction to the Political Theory of Roberto Esposito*. London, Routledge, 2015.

Brown, Wendy. *Walled States, Waning Sovereignty*. New York, Zone Books, 2010.

Butt, Amy, "Held in Common: Science Fiction and Collective Space." Un/Building the Future: The Country and the City in the Anthropocene, 14–16 June 2023, University of Warwick, Keynote Address.

Dean-Jones, Lesley. “Menstrual Bleeding According to the Hippocratics and Aristotle.” *Transactions of the American Philological Association (1974–2014)*, vol. 119, 1989, pp. 177–91, https://doi.org/10.2307/284268.

Edmond, Rod, and Vanessa Smith. “Editor’s Introduction.” *Islands in History and Representation*, edited by Rod Edmond and Vanessa Smith. Routledge, 2003, pp. 1–18.

Franks, Jill. *Islands and the Modernists: The Allure of Isolation in Art, Literature and Science*. Jefferson, NC, McFarland, 11 July 2006.

Gillis, John R. “Taking History Offshore: Atlantic Islands in European Minds, 1400–1800.” *Islands in History and Representation*, New York, Routledge, 2003, pp. 19–31.

Goldstein, Joshua S., and Devid P. Rapkin. “After Insularity: Hegemony and the Future World Order.” *Futures*, vol. 23, no. 9, 1991, pp. 935–59.

Groneman, Carol. “Nymphomania: The Historical Construction of Female Sexuality.” *Signs: Journal of Women in Culture and Society*, vol. 19, no. 2, 1994, pp. 337–67.

Grosz, Elizabeth. *Volatile Bodies: Toward a Corporeal Feminism*. Bloomington, Indiana UP, 1994.

Hayward, Eva. “Transxenoestrogenesis.” *TSQ: Transgender Studies Quarterly*, vol. 1, no. 1–2, 1 January 2014, pp. 255–58, https://doi.org/10.1215/23289252-2400190. Accessed 9 September 2021.

Hofmeyr, Isabel. “Oceans as Empty Spaces? Redrafting Our Knowledge by Dropping the Colonial Lens.” *The Conversation*, 6 September 2018, theconversation.com/oceans-as-empty-spaces-redrafting-our-knowledge-by-dropping-the-colonial-lens-102778#:~:text=The%20sea%20is%20a%20flat. Accessed 14 November 2023.

Hui, Sylvia. “Report: UK Clearly Overwhelmed by Surge in Migrant Crossings.” *AP News*, 21 July 2022, https://apnews.com/article/immigration-migration-english-channel-88ba5e92a832660f6fb59b9eb51ba301. Accessed 14 November 2023.

Klaiber, M. *Insularity vs. Islomania: The Island Setting and Shipwreck Experience in Three Dystopian Novels*. Munich, Grin Verlag, 2023.

Mackintosh, Sophie. *The Water Cure*. Kindle ed., New York: Doubleday, 2019.

MacLeod, Janine. “Holding Water in Times of Hydrophobia.” *Petrocultures: Oil, Politics, Culture*, edited by Sheena Wilson et al., Montréal & Kingston; Ithaca, NY, Mcgill-Queen’s University Press, 2017, pp. 264–86.

McMahon, Elizabeth. *Islands, Identity and the Literary Imagination*. London, Anthem Press, 2016.

Meeker, Joseph W. “Subtleties of the Isle: Islands and the Imagination.” *Interdisciplinary Studies in Literature and Environment*, vol. 18, no. 1, 23 March 2011, pp. 197–202.

Miller, Sarah Alison. *Virgins, Mothers, Monsters: Late-medieval Readings of the Female Body out of Bounds*. University of North Carolina Digital Repository 2008. https://doi.org/10.17615/n3g4-tk64

Moylan, Tom. *Scraps of the Untainted Sky*. London, Routledge, 2020.

Neimanis, Astrida. *Bodies of Water: Posthuman Feminist Phenomenology*. London, Bloomsbury Academic, 2017.

Tsing, Anna Lowenhaupt. *Mushroom at the End of the World: On the Possibility of Life in Capitalist Ruins*. Princeton, NJ, Princeton University Press, 2015.

Part 3

More-Than-Human Mutual Aid and Eco-Justice

8 Dystopian Prohibitions and Utopian Possibilities in Edmonton, Canada, at the Onset of the COVID-19 Pandemic

Rylan Kafara

Introduction

Gentrification continues worldwide in the COVID-19 pandemic. Edmonton's gentrification process centers on Rogers Place, which is the new home for the National Hockey League's (NHL) Edmonton Oilers, the ice hockey franchise owned by billionaire Daryl Katz. The arena's 613.7 million Canadian dollars (CAD) building cost was debt-financed by the City of Edmonton (COE). In an agreement with the COE, The Katz Group receives all revenue from the publicly funded arena until the 2050s (Scherer et al., *Power Play* 303, 310). This revenue includes profiting from the arena's naming rights, purchased by Rogers Media (Canadian Press). The Katz Group also re-named the gentrifying area the Ice District, which continues to be reshaped with premium office spaces, upscale condominiums, restaurants, and hotel services catering to wealthier residents and tourists. Existing social housing, working class pubs, and more affordable leisure spaces were removed (Scherer et al., "Negotiating" 112).

Gentrification in Edmonton was briefly halted at the onset COVID-19. In March 2020, the NHL paused its season, and games at Rogers Place were cancelled or delayed. Many Edmontonians sheltered in place under early COVID-19 social distancing guidelines. The federal branch of the settler-state transferred billions in pandemic relief funding to the municipal and provincial branches (Johnson). On average, Albertans received the most funding of all Canadian provinces, including CAD 1.9 billion in business loan forgiveness (CTV). However, over CAD 675 million of federal pandemic relief directed to the provincial government was unused in Alberta. These unspent funds were designated for low-wage workers, housing, and long-term care (Johnson). Municipal leadership chose to use the last of its COVID-19 funding to create a CAD 22.9 million tax break program for downtown developers instead of "helping vulnerable populations" (Boothby; also see Cook). This chapter argues that, during COVID-19, the gentrification of city-center Edmonton, within the wider structure of racial capitalist settler coloniality, continues to be prioritized over residents most at risk from ongoing structural harms exacerbated during the pandemic. These residents include Edmonton's houseless community, the majority of whom are Indigenous (Homeward Trust).

DOI: 10.4324/9781003345770-12

Continued capital accumulation in Edmonton during the pandemic is a coordinate on the continuity of settler time, which Mark Rifkin argues is the linear spatiotemporality structuring of the United States in the past, present, and future (*Beyond* 2). Rikfin shows how

> U.S. settler colonialism produces its own temporal formation, with its own particular ways of apprehending time, and the state's policies, mappings, and imperatives generate the frame of reference such as plotting events with respect to their place in national history and seeing change in terms of forms of American progress. (2)

Similarly, Canadian settler colonialism also produces its own particular temporality. I suggest Edmonton, a Canadian prairie city, is structured by a local variant of settler time. This chapter concentrates on the early months of the COVID-19 pandemic and explains how the pandemic heightened existing spatial and temporal boundaries. These pandemic boundaries created both dystopian prohibitions, including restriction, containment, and exclusion, and utopian possibilities, including harm-reduction, mutual care, and reciprocity. The dystopian prohibitions protected Hub City: a fenced off area of city-center Edmonton enclosed to host the 2020 NHL Playoffs and Stanley Cup final.[1] Hub City continued global capital accumulation amidst the disaster of the pandemic.[2]

Contrasting Hub City was Pekiwewin, an Indigenous-led, unsanctioned encampment where roughly 400 houseless Edmontonians resided.[3] Pekiwewin offered utopian possibilities for living concurrently to racial capitalist settler coloniality's extractive and transactional relationships. Pekiwewin made time and space for relationships of reciprocity and mutual care. Instead of only carving out recognition (Coulthard 3) within racial capitalist settler coloniality, Pekiwewin was in an Indigenous temporality concurrent to settler time. Such a temporality offers hope for imagining futurities outside new pipelines of racial capitalism.

Settler Time in City-Center Edmonton

Currently, actual pipelines carve out capitalist extractions to connect the region and its resources to international markets. Inefficient and extractive, capitalism requires expansion by exploiting new uses of land and labor. This expansion, as Rhiannon Firth explains, occurs in "an ever-shrinking frontier" (60). Naomi Klein traces how, since the beginning of the neoliberal era, a new route, or pipeline, of continued accumulation has been capitalizing on disaster. In Edmonton, a key and consistent form of capital accumulation is iterations of Indigenous dispossession. As the COVID-19 pandemic began in Edmonton, a new pipeline of disaster capitalism was built and coalesced with two ongoing crises of capital accumulation: the disasters of the drug poisoning crisis and the housing crisis.

Since the onset of the pandemic, Edmonton's city-center residents—or those houseless, sleeping rough, or precariously housed—were mostly Indigenous. Since settlement in the 1870s, Edmonton's city-center was framed as a site of social disorder by business and political elites and their aligned media pundits, together known as "boosters" (Hicks; Scherer et al., *Power Play* 310–12, 320–22). The city-center's underdeveloped land, coupled with visibility of Indigenous and other diverse communities, was primed for revanchist gentrification (Smith). Revanchism, as Granzow and Dean explain, is a circular form of vengeance directed against houseless Indigenous Peoples in Edmonton for not disappearing when settlers steal their land (103). Gentrification, broadly, reshapes land and its uses for capital accumulation, resulting in the displacement and dispossession of pre-existing residents through architecture, evictions, and policing (Kafara 291–93). Gentrification is heralded as a transformative solution to social disorder through the revanchist resettlement of city-center Edmonton primarily by white, wealthier settlers. Amid this resettlement, houseless Edmontonians constantly keep moving.

Scherer et al.'s ethnography of the Ice District from 2016 to 2018 shows how city-center Edmonton continues to be structured by settler colonialism ("Negotiating"), which is an iterative, ongoing process of controlling territory through Indigenous dispossession, erasure, and elimination (Wolfe) aligned with racial capitalism, or the racialization of capitalism's extractive relationality with land and labor (Robinson). Despite settler colonialism continuing, its harms are relegated to the past by the modern Canadian settler-colonial state. As Conrad Scott shows, Indigenous dystopian writers know these harms did not end, but are a "continued catastrophe" (12). Granzow argues that in Edmonton, these continued harms are ignored by an indifferent settler society.

Following Rogers Place opening in 2016, those sanctioned to participate in the arena economy—such as hockey fans and concert attendees—moved freely throughout the Ice District, while city-center residents were displaced through policing. This is indicative of ongoing settler colonialism. It normalizes white entitlement to Indigenous land and resources and erases the bodies and histories of predominantly Indigenous city-center residents (Scherer et al., "Negotiating").

Edmonton is a settler territory comprised of urban space formed from dispossessed Indigenous land that has been privatized and commodified. This process is normalized in the settler mindset through what Rifkin calls "settler common sense," or "the ways the legal and political structures that enable non-native access to Indigenous territories comes to be lived as given, as simply the unmarked, generic conditions of possibility for occupancy, association, history, and personhood" (*Settler* xvi). As cities grew, *terra nullius* became *urbs nullius*, or "urban space devoid of Indigenous sovereign presence" (Coulthard 176). The ideology of *urbs nullius* continues the erasure of Indigenous presence in settler territory, specifically in growing urban spaces like Edmonton, where any uses of land outside of racial capitalism's forms of capital accumulation is unrecognized or controlled.

City-center Edmonton's spatiotemporality is on what Rifkin calls "the grid of homogenized space and time" (*Beyond* 24). On this grid, Rifkin explains, there is an "ostensible copresence in space (sizing that grid at whatever scale—a particular region, the territory of the nation-state, the globe) on a slice of time (however wide—an instant, a year, a decade)" (23). Just as settler colonialism failed to eliminate Indigenous Peoples, settler time is not "fully able to displace Indigenous temporal orientations" (2). Understanding Edmonton's variant of settler time helps reveal the city's development, its spatiotemporality early in the COVID-19 pandemic, and its continued trajectory in the Canadian settler-state.

Racial capitalism in Edmonton is reproduced by its settler spatiotemporality. As Edmonton formed, so did its variant of "institutionalized authority over 'domestic' territory," which "powerfully shapes the possibilities for interaction, development, and regularity within it" (*Beyond* 2). This is Edmonton's spatiotemporality, and erasure of an Indigenous co-presence is a critical aspect of this structure. Settler time in Edmonton exists along a singular trajectory of particular coordinates—a before, a during, and an after—enabling a spatiotemporal continuity by neatly separating the past, present, and future. Normalizing iterations of settler colonialism frames initially stealing land, the Indian Residential School (IRS) system, the pass system, and stealing children in the Sixties Scoop, as all being dealt with and "settled" (Simpson 11; also see Maynard and Simpson). I suggest this partitioning is fundamental to settler time continuing. This settler trajectory also structures Edmonton's present: the contemporary foster care system, carceral system, and ongoing forms of displacement and dispossession of Indigenous Peoples have ostensibly been disconnected from the settler-state's previous modes of genocide, thereby enabling their separation in the indifferent settler mentality. This also helped make the momentary dystopia of Hub City seem not only normal, but something to applaud. New, consumptive entertainment was prioritized and produced despite the resources and risks it required.

Crucial to the continuity of settler time was the framing of the Oilers' move into Rogers Place. The new home of the Oilers was celebrated with the unveiling of Saulteaux artist Alex Janvier's *Iron Foot Place* in Rogers Place. The mediated framing of this act of "settler reconciliation" (Davidson et al.) obscures the ongoing genocidal process of Indigenous dispossession, erasure, and elimination. Excluded from the mediated framing was that the opening of Rogers Place resulted in further Indigenous dispossession and displacement in city-center Edmonton (Scherer et al., "Negotiating").

Throughout this linear trajectory of settler time in Edmonton, Indigenous Peoples are portrayed as either existing in the past and dysfunctional in modernity, or succeeding within settler society by departing from a primitive existence and adopting settler interests and values such as sports fandom and capital accumulation. Settler time, then, ignores and obscures ongoing Indigenous refusal to disappear. Gerald Vizenor calls this refusal "survivance" ("Aesthetics of Survivance" 1). Survivance, Vizenor explains, is an "active presence" (*Fugitive Poses* 15). From Indigenous children playing ice hockey in the IRS system to Janvier's *Iron Foot Place*, survivance "[i]s more than survival, more than endurance or mere response" (15), but instead, "active repudiation of dominance, tragedy, and victimry" (15). As Audra Simpson asserts, settler colonialism "fails at what it is supposed to do: eliminate Indigenous people; take all their land; absorb them into a white, property-owning body politic" (7–8). Settler colonialism was, and continues to be, a failure.

Although settler colonialism has not succeeded, it continues to be mutable and insidious in achieving its goal of elimination. Settler time is "an ongoing colonial imposition that denies Indigenous peoples' histories, sovereignties, and self-determination" (Rifkin, *Beyond* viii). I suggest that Edmonton's spatiotemporality—its orientation, trajectory, and continuity—is fixed on settler time and continues this colonial imposition on Indigenous Peoples. As Rifkin explains, "Indigenous and settler temporalities (and associated ways of imagining and occupying place) can be treated as co-occurring, but in ways that do not make them part of an inherently singular present" (*Settler* 35). This was how Hub City and Pekiwewin were two different, but concomitant, spatiotemporalities. So, rather than one singular settler time, or one "shared, unified 'now' (which includes a shared 'then' of the past)" (Rifkin, *Beyond* 1), there was a "temporal multiplicity" (16) with "multiple processes of becoming that have their own trajectories" (17). Drawing from an ethnography of city-center Edmonton during both Hub City and Pekiwewin, what follows is an exploration of the co-presence of pluralistic and reciprocal futures.

The Dystopian Prohibitions of Hub City and the Utopian Possibilities of Pekiwewin

Describing the prohibitions and mediated framing of Hub City helps imagine a dystopian future. Creating and sustaining Hub City ensured the safety, health, and leisure of those inside its fencing instead of supporting those more at risk from the harms of the pandemic and the ongoing housing and drug poisoning disasters. The protection of Hub City required the deployment of police and racialized workers, plus the settler-state spending millions (Progress Report). Fencing covered in Ice District branding was installed across Hub City's perimeter, encasing Rogers Place and nearby upscale hotels and restaurants in the professional sports bubble. Hub City enclosed teams, coaching and support staff, and broadcasters from across the continent. Additionally, local workers tended to bars, served in restaurants, and supported media outlets. Health workers conducted roughly 1,500 daily tests of everyone inside Hub City (Spencer).

Although the safety, health, and leisure of those inside Hub City was sustained throughout the playoffs, it was achieved through prohibition. Players and team staff could not leave Hub City or have any visitors. The workers were separated from their loved ones for their safety. This was predicated upon workers fulfilling their role in maintaining what Adele Pavlidis and David Rowe call "the mediated sporting spectacle under the pandemic." Hub City's professional sport experience was mediated to viewers consuming broadcasted games. The fencing protecting the spectacle was guarded day and night by contract security guards. These guards lacked training and were primarily staffed by newcomers from Asia and Africa (field notes, 17 September 2020). Couriers working in the gig economy dropped food off at the gates, while Edmonton's police force directed traffic.

To obscure the reality of Hub City, its immediate perimeter was excluded in its mediated framing. For instance, the Ice District branded covers fell off Hub City's fencing, but fans watching the broadcasts of games never saw the derelict condition of the perimeter. Instead, viewers saw aerial views of the nearby river valley, or the Rockies, a mountain range hundreds of kilometers away. Depictions of city-center Edmonton included the incursion of war planes flying over Rogers Place on 25 September 2020 (field notes). The flyover of two jet fighters was timed with the singing of the Canadian National Anthem inside Rogers Place (Sportsnet). This mediated framing was synchronous with Rifkin's settler time, exemplified in Edmonton's variant: "notions, narratives, and experiences of temporality that de facto normalize non-native presence, influence, and occupation" (*Beyond* 9). Both the militarized and bordering aspects of the Canadian settler-state were exemplified in this spectacle.

The same moment in the pandemic also revealed the utopian possibilities of an Indigenous temporal co-presence to settler time. Contrasting the restrictive prohibitions and mediated spectacle of Hub City was the harm-reduction, mutual care, and reciprocity of Pekiwewin. As the perimeter of Hub City was established, a tipi was raised nearby in the river valley. Pekiwewin began with the raising of the tipi and the lighting of the sacred fire on the morning of 24 July 2020. Land used as an overflow parking lot for the adjacent baseball stadium became space outside of settler-colonial displacement, dispossession, and erasure. Instead, Pekiwewin was—both relationally and temporally—nested Indigenous sovereign territory, or what Simpson refers to as Indigenous territory "within and apart from settler governance" (11). This remaking was not a return to the past, but the "re-creation" of something new not just outside of settler territory, but also settler time (Rifkin, *Beyond* 32). "Indigenous knowledge," as Adam Barker and Jenny Pickerill explain, "practiced in relation to land and place, offers a necessary challenge to settler colonial values by espousing mutual care, obligation and reciprocal relations of responsibility" (1706–07). These relations are crucial for a sustainable future.

At the onset of the COVID-19 pandemic, the present was grim. By 14 March 2020, transmission of COVID-19 was identified in Edmonton (Wakefield). During the gentrification of Ice District before the pandemic, city-center residents were increasingly concentrated into the non-profit industrial complex (NPIC). The NPIC regulated where, and when, city-center residents could be without being moved along by police. Under the guise of neoliberal efficiencies, the NPIC was comprised of nongovernmental organizations (NGOs). These NGOs were officially at arm's-length from the settler-state to keep their workers underfunded, under-resourced, depoliticized, and non-unionized while ensuring the continuity of racial capitalist settler coloniality (Incite! Women of Color Against Violence). Amid the uncertainty of the early pandemic, city-center residents were further surveilled, encased, and displaced (interview, 27 January 2021). NPIC support centers either had their capacity limited or were closed. Supports for city-center residents were concentrated into one main daytime shelter in an events center called EXPO, and one main overnight shelter at a recreation facility called the Kinsmen Sports Centre.

NPIC frontline harm-reduction workers were concerned with the lack of supports for city-center residents. Outreach workers continued to witness violent encampment evictions, even during the COE's supposed moratorium of policing and destroying encampments (interview, 27 January 2021). Contradictory information was mis-communicated to city-center residents. Pre-pandemic inconsistency was routine in the government and NPIC's treatment of houseless Edmontonians. As the Kinsmen and the EXPO centers were closing in summer 2020—with no wind-down plan or alterative COVID-19 supports in place for city-center residents—many felt it was time to take action. So, in June, a meeting was held. Attendees included Indigenous and non-Indigenous frontline harm-reduction workers, community activists, and relations of unhoused Edmontonians. Together, through sharing resources and relationships, the "action of care" that became Pekiwewin began to take shape (field notes, 17 June 2020).

Relationships were fundamental in this planning. Two Indigenous co-organizers, Deanna and Veronica, had friends in other cities with experience in operating encampments. In Edmonton at the time, Deanna recalls, "people didn't really know how to keep each other safe. … If the government's not going to keep us safe, who will?" (interview, 7 July 2021). Drawing on those friendships, Veronica remembers, in order to create Pekiwewin, it took "the nerve to think that this was possible" (interview, 26 July 2021). In the immediacy of the early pandemic, this nerve was used to take action.

Intent and relationships created, sustained, and maintained Pekiwewin, even during Pekiwewin's closure. These relationships helped grow the encampment organically. As word spread, city-center residents arrived to camp, supporters dropped off supplies, and volunteers worked in the provision of care. A sign beside the tipi displayed "Pekiwewin Camp is not a protest camp. Pekiwewin is an anti-police violence camp with a harm reduction approach for houseless/people sleeping rough led by Indigenous 2spirit women and femme folks working in solidarity with Black, LGBTQ2S & settler allies!" (field notes, 27 July 2020). Another sign said "you are loved, you are welcome" (field notes, 13 August 2020). Pekiwewin was not a protest, but an action of care.

Also explicit was the banner displayed at the south end of camp in red lettering: "THIS IS NATIVE LAND" (field notes, 2 September 2020), proclaiming Pekiwewin on a concurrent trajectory to settler time. Rifkin explores Indigenous "ways of being-in-time that are not reducible to participation in a singular, given time—a unitary flow—largely contoured by non-native patterns and priorities" (*Beyond* 3). The spatiotemporality of city-center Edmonton and, specifically, Hub City, kept houseless residents moving—even when Edmontonians were expected to shelter in place. This spatiotemporality could not fully permeate Pekiwewin's borders, allowing Pekiwewin residents respite as long as they pleased. There was a food tent that served meals and snacks. A community resource space provided first-aid, drug poisoning reversals, and harm-reduction supplies. There was a tent with clothing, a library with books, art supplies, and Wi-Fi, and a workout area with weights. Crucially, there was sanitation, including outhouses and garbage bins.

Pekiwewin was the inverse of Hub City. Police were not permitted at Pekiwewin. After discussions between the COE and Pekiwewin organizers, NPIC workers were allowed in to build relationships with camp residents and work on long-term supports such as housing. Importantly, NPIC workers operated under the guidance of Pekiwewin volunteers. Moreover, the COE promised to handle waste removal for the outhouses and garbage. These were key practices in Pekiwewin's sustainability and an outcome of the work of its organizers. Through these practices, Pekiwewin circumvented what Firth calls "recuperation" (4). Edmonton's variant of recuperation occurred through the non-profitization/NGO-ization of resistance to racial capitalism. Pekiwewin was not a resistance to racial capitalist settler coloniality, but a refusal. By staying visible, Pekiwewin was not recuperated into the NPIC, and avoided being subordinated to the settler-state, its response to the disaster of the pandemic, and its handling of the housing and drug poisoning crises.

Throughout camp's duration from 24 July to 12 November 2020, everyday practices of mutual care were shared between camp organizers, residents, and volunteers. Communal events like karaoke and film screenings were organized. Through intentional and caring relationships, these mutual care practices were grounded in reciprocity. For example, camp residents taught inexperienced volunteers how to conduct drug poisoning reversals (field notes, 20 August 2020). These volunteers then dealt with subsequent drug poisonings, which allowed camp residents respite from routinely reversing drug poisonings, which are an everyday occurrence in city-center Edmonton because of an unsafe supply and lack of regulation of drugs by the settler-state. In turn, for many Indigenous and non-Indigenous people, drug use is a response to harms caused by racial capitalist settler coloniality. Referencing the generational harms throughout settler time in Edmonton, from the early forts, through the IRS system and the pass system, to the foster care system and incarceration, an Indigenous camp resident named Jenny explained that "the healing of Indigenous people is stunted, and we treat it with drugs" (field notes, 29 October 2020).

Pekiwewin was in a spatiotemporal co-presence to settler time that created and sustained a space and time for people to build reciprocal relationships and to heal. For a moment, it helped to imagine the utopian possibilities of reciprocity and healing, processes that are stunted amidst an indifferent Edmonton settler society (Granzow). This moment was not sustained at Pekiwewin due to the incursion of settler time, as its capacity to maintain its borders was reduced. This occurred through the COE unreliably conducting garbage pickups, the police dropping off houseless residents in crisis at the edges of the site, the mediated framing of Pekiwewin as a site of social disorder, and the everyday violence of city-center Edmonton—resulting from the informal capitalist economy and gang involvement—infringing overtly into Pekiwewin. These incursions cannot be used to dismiss how Pekiwewin gave residents space and time to heal that they did not have earlier, anywhere else at the time, or afterward.

Due to the relentlessness of these incursions, organizers decided to wind Pekiwewin down in mid-November 2020. Cold weather was also an issue but, by that time, the camp was winterized. The constant incursions, however, made "livable forms of stability" (Rifkin, *Beyond* 32) untenable, at least in the short term. So, following a meeting, including Indigenous Elders, organizers, and volunteers, the camp's closure began with a sharing circle and letting-go ceremony. The tipi was taken down and the sacred fire put out. On 7 November, a solidarity rally was held at the nearby Alberta Legislature followed by a ceremonial round dance beside the encampment. Pekiwewin's presence forced the COE to make a last-minute plan for supporting city-center residents during winter amidst the ongoing pandemic (field notes, 18 November 2020). The Edmonton Convention Centre was hurriedly transformed into Tipinawâw (Nêhiyawak/Cree for it is sheltered), a 24-hour winter shelter funded by the COE and operated by the settler-led NPIC from the end of October 2020 to the end of April 2021.

Despite Tipinawâw opening, some camp residents stayed at Pekiwewin. A team of volunteers stayed to support camp residents through the inevitable police eviction, which occurred on 12 November 2020. Over two dozen police officers supervised the eviction, conducted with peace officers, city workers in hazmat suits, and contractors from a fence rental company. Camp residents were removed, and fencing was installed around the site's perimeter. The fencing stayed put until the site was once again an overflow parking lot. Any city-center residents entering the site experienced an immediate police response and had to keep moving.

The settler-state's actions during the pandemic offer a glimpse into Edmonton's future. If settler time continues along its linear path of accumulation through dispossession, Edmonton's frontier will run out of new routes of extraction and contract into a dystopian past when settlers took refuge in a fort.[4] Those outside the fencing in this future will be left to die. Concurrently, Pekiwewin showed a futurity more livable and more sustainable for both Indigenous and non-Indigenous people.

Coda: The Prohibitions and Possibilities in Edmonton's Present

Today, racial capitalism's frontier expands in tandem with rising drug poisoning deaths and the visibility of houseless Edmontonians. Alberta's Provincial government, under consequent Premiers Jason Kenney and Danielle Smith, has decimated harm-reduction initiatives through prohibition policies since the onset of COVID-19. The government closed supervised consumption sites (SCS), funded private forced recovery programs, and deliberately hid the number of drug poisoning deaths from the public (Rusnell; Thomson). Data released following Smith's victory in the 29 May 2023 election revealed April 2023 had the highest number of drug poisoning deaths ever recorded in Alberta (Kanygin). Thomson et al. suggest that "by accelerating drug toxicity harms disproportionately borne by Indigenous people, drug prohibition legislates Indigenous genocide." These policies are among the settler-state's latest dystopian prohibitions.

In city-center Edmonton, drug prohibition is coupled with a punitive response to the increased visibility of houselessness and people who use drugs in a mediated framing of social disorder: the settler-state deploys more police. The growing carceral system imprisons Indigenous Peoples at a much higher rate than other residents. Although Indigenous Peoples only comprise 5.5% of Edmonton's total population, they account for more than 65% of imprisoned women, and nearly 60% of imprisoned men in Edmonton (Scherer et al., "Spectacle"). Drug prohibition is a pipeline to disaster: since the onset of the drug poisoning crisis, the life expectancy of Indigenous Peoples in Alberta dropped seven years (Paradis). Indigenous Peoples are also seven times more at risk than settlers to die of a drug poisoning in Alberta (Sousa).

As it does across the world, displacement and dispossession of Indigenous Peoples continues in the gentrifying Ice District. This continuity is accompanied by both Indigenous survivance and recognition in ongoing racial capitalist settler coloniality. In late 2021, a harm-reduction center adjacent to Rogers Place was purchased by the Katz Group. The center, operated by NPIC agency Boyle Street Community Services (BSCS), is one of the only spaces houseless Edmontonians can find refuge during the winter. To move, BSCS received a large donation from the Katz Group's charity, the Edmonton Oilers Community Foundation, which raises money from Oilers fans at games and other events. The donation helped BSCS purchase a new building further away from the Ice District. Recuperation remains crucial to gentrifying the city-center. As Scherer et al. note, moving BSCS was key since the initial plan for Ice District in 2008 ("Dirt City" 507), but BSCS secretly bought the center's property in 2013, the year the Rogers Place arena agreement was finalized between the COE and the Katz Group. Although displacement was delayed, the Katz Group not only finally achieved its goal, but was praised in the move's mediated framing. This impending move exacerbates risks of harm. It further concentrates the NPIC into a smaller area and has already fractured long-term community relationships that had collectively resisted the harms of gentrification catalyzed by Rogers Place. Hopefully, these relationships can be mended. On 12 September 2023, BSCS announced its lease with the Katz Group was ending 30 September (Mulcahy). The center was closed and its supports dispersed, causing immediate upheaval, harm, and death (personal correspondence, 6 October 2023). As I write this, BSCS' new building is not ready, and winter is coming.

Although the future is uncertain, Kelly Hayes reminds us that, "in a world that is breaking down our connections, isolating us, and sub-siloing us to death, life-giving relationships are our best hope" (222). The relationships that created and sustained Hub City were transactional, extractive, and exploitative. These relationships serviced the health of those inside and ensured continued accumulation of capital in global racial capitalism amid the pandemic. Through the mediated spectacle of the NHL playoffs, Hub City was affixed to a pandemic coordinate on Rifkin's "grid of homogenized space and time" (*Beyond* 24) that was both a return to the past and a suggestion of a dystopian future.

The spatiotemporal co-presence of Pekiwewin not only juxtaposed the relationships of Hub City but also gave hope for a sustainable future grounded in life-giving relationships. As bergman et al. document, this hope was widespread: from the onset of the pandemic, Indigenous Peoples conducted mutual aid across the continent (185). Pekiwewin did not mark a return to a primitive past, but rather revealed an ongoing Indigenous temporal co-presence with settler time. This co-presence was grounded in connection to land that was a meeting place for Indigenous Peoples for thousands of years. After Pekiwewin's closure, much of its work and many of its workers were recuperated into the NPIC. Giving Indigenous names to settler-led NPIC centers and initiatives has become a common practice. Pekiwewin leadership and volunteers were hired into precarious frontline harm-reduction jobs. Tipinawâw was used by the provincial government to shut down Edmonton's busiest SCS at BSCS and redeploy the harm-reduction staff to the shelter. When Tipinawâw closed, BSCS' SCS never reopened.

Concurrently, mutual aid groups that participated in Pekiwewin expanded to maintain relationships with former Pekiwewin residents. The risk of this mutual aid work, however, is the settler-state using these groups as it uses the NPIC. As Firth argues, the state's reliance on community aid at the beginning of COVID-19 was not unique to the pandemic, but part of a larger, global pattern of the state's "incapacity and indifference" (6). As settler-state resources were funneled into corporations like the Katz Group, Rogers Media, and the NHL, the provision of care continued to be privatized.

Hope remains, however, through mutual care practices—in particular reciprocal, harm-reduction practices conducted by Indigenous-led autonomous collectives—that have not been recuperated by the settler-state. Pekiwewin's closure did not end the Indigenous temporal co-presence that racial capitalist settler coloniality has failed to erase. This gives both Indigenous and non-Indigenous Peoples hope for a sustainable future. Instead of fences, more tipis—or other forms signifying an Indigenous spatiotemporal co-presence to settler time—will be raised again in city-center Edmonton.

Acknowledgments

Thanks to Fred Mills and Tony Bolger for providing feedback on earlier chapter drafts. Thanks also to my PhD supervisors Judy Davidson and Jay Scherer, as well as Paul Aya, Izabela Bienko, Blake Hamer, Suzanne Lenon, Kalie Stieda, Andrea Taylor, and Dan Zimmerman for participating in Hub City walks. Finally, thanks especially to Deanna Cachagee, Piyêsiw Crane, Veronica Fuentes, Kaela Siewert, and everyone who participated in creating and sustaining Pekiwewin, and continuing its relationships since its closure. This research was possible through support from the Social Sciences and Humanities Research Council and the Killam Trusts.

Notes

1 During the 2020 playoffs, all players, coaches, and team staff lived inside Hub City. If anyone was caught breaching the fencing, they were sent home (field notes, 5 August 2020).

2 This was reminiscent of a series of five trading posts called Fort Edmonton (1795–1915) that connected Edmonton to the global fur trade.

3 This is my first major publication drawing from dissertation fieldwork conducted during Pekiwewin amid a multi-year ethnography of city-center Edmonton. Future publications, my dissertation, and other research outputs will be available at www.CityCentreEdmonton.com.

4 Settlers used Fort Edmonton in 1885 for refuge, fearing an Indigenous attack that never happened (Goyette and Roemmich, 100).

Works Cited

Barker, Adam J., and Jenny Pickerill. "Radicalizing Relationships to and through Shared Geographies: Why Anarchists Need to Understand Indigenous Connections to Land and Place." *Antipode*, Vol. 44, No. 5, 2012, pp. 1705–25, DOI: 10.1111/j.1467-8330.2012.01031.x.

bergman, carla, magali rabasa, Ariella Patchen and Seyma Özdemir. "Turtle Island." *Pandemic Solidarity: Mutual Aid during the Covid-19 Crisis*, edited by Marina Sitrin and Colectiva Sembrar, Pluto Press, 2020, pp. 183–229.

Boothby, Lauren. "City Eyes Possible $23 Million in Grants for Big Downtown Projects Starting Construction to Offset COVID-19 Impacts." *Edmonton Journal*, 19 January 2021, https://edmontonjournal.com/news/local-news/city-eyes-possible-23-million-in-grants-for-big-downtown-projects-starting-construction-to-offset-covid-19-impacts

Canadian Press. "Rogers Purchases Naming Rights for Oilers' Arena," *Sportsnet*. 3 December, 2013, https://www.sportsnet.ca/hockey/nhl/rogers-purchases-naming-rights-for-oilers-arena/.

Cook, Dustin. "Downtown will be Back:" Edmonton City Council Approves $22.9-Million Tax Break Grant Program for Residential Developers in the Core." *Edmonton Journal*, 25 January 2021, https://edmontonjournal.com/news/local-news/downtown-will-be-back-edmonton-city-council-approves-22-9-million-tax-break-grant-program-for-residential-developers-in-the-core.

Coulthard, Glen. *Red Skin, White Masks: Rejecting the Colonial Politics of Recognition*. U of Minnesota P, 2014.

CTV News Edmonton. "Albertans Averaged More Federal COVID-19 Financial Support Than Any Other Province: Report," *CTV News*, 26 August 2021, https://edmonton.ctvnews.ca/albertans-averaged-more-federal-covid-19-financial-support-than-any-other-province-report-1.5562734.

Davidson, Judy, Jay Scherer, and Rylan Kafara. "Alex Janvier's Iron Foot Place: Indigenous Public Art, Survivance, and Settler Reconciliation in Edmonton's Rogers Place." *Journal of Canadian Studies*, Vol. 56, No. 2, 2022, pp. 468–94, DOI: 10.3138/jcs-2021-0017.

Firth, Rhiannon. *Disaster Anarchy: Mutual Aid and Radical Action*. Pluto Press, 2022.

Goyette, Linda and Carolina Jakeway Roemmich. *Edmonton: In Our Own Words*. U of Alberta P, 2004.

Granzow, Kara. *Invested Indifference: How Violence Persists in Settler Colonial Society*. UBC Press, 2020.

Granzow, Kara and April Dean. "Revanchism in the Canadian West: Gentrification and Resettlement in a Prairie City." *Topia*, Vol. 18, 2007, pp. 89–106, DOI: 10.3138/topia.18.89

Hayes, Kelly and Mariame Kaba. *Let This Radicalize You: Organizing and the Revolution of Reciprocal Care*. Haymarket Books, 2023.

Hicks, Graham. "Rogers Place Helping Edmonton Take Back the Night." *The Edmonton Sun*, 14 October 2016, https://edmontonsun.com/2016/10/14/rogers-place-helping-edmontontake-back-the-night.

Homeward Trust. "Number of People in Edmonton in 2023." *Homeward Trust*, 24 February 2023, https://homewardtrust.ca/data-analytics-reporting/.

Incite! Women of Color against Violence. *The Revolution Will Not Be Funded: Beyond the Non-Profit Industrial Complex*. South End Press, 2007.

Johnson, Lisa. "Alberta Leaving More Than $675 Million in Federal Emergency Funds Unspent: Report." *Edmonton Journal*, 27 January 2021, https://edmontonjournal.com/news/politics/alberta-leaving-more-than-675-million-in-emergency-funds-unspent-report.

Kafara, Rylan. "'Who Is Really Gonna Benefit?': The Punk Habitus in the Downtown Edmonton Field." *Punk & Post-Punk*, Vol. 9, No. 2, 2020, pp. 287–307, DOI: 10.1386/punk_00029_1.

Kanygin, Jordan. "Alberta Marks Deadliest Month for Overdose Deaths in April." *CTV News*, 23 June 2023, https://calgary.ctvnews.ca/alberta-marks-deadliest-month-for-overdose-deaths-in-april-1.6457822.

Klein, Naomi. *The Shock Doctrine: The Rise of Disaster Capitalism*. Picador, 2007.

Maynard, Robyn, and Leanne Betasamosake Simpson. *Rehearsals for Living*. Alfred A. Knopf Canada, 2022.

Mulcahy, Karyn. "Boyle Street Community Services Closing Downtown Location at the End of September." *CTV News,* 12 September 2023, https://edmonton.ctvnews.ca/boyle-street-community-services-closing-downtown-location-at-the-end-of-september-1.6558216.

Paradis, Danielle. "First Nations Life Expectancy Plummets in Alberta due to Opioid Deaths." *APTN News,* 4 July 2023, https://www.aptnnews.ca/national-news/first-nations-life-expectancy-plummets-in-alberta-due-to-opioid-deaths/.

Pavlidis, Adele, and David Rowe. "The Sporting Bubble as Gilded Cage: Gendered Professional Sport in Pandemic Times and Beyond." *M/C Journal*, Vol. 24, No. 1, 2021, DOI: 10.5204/mcj.2736.

Rifkin, Mark. *Beyond Settler Time: Temporal Sovereignty and Indigenous Self-Determination.* Duke U P, 2017.

Rifkin, Mark. *Settler Common Sense: Queerness and Everyday Colonialism in the American Renaissance.* U of Minnesota P, 2014.

Robinson, Cedric. *Black Marxism: The Making of the Black Radical Tradition.* Zed Press, 1983.

Rusnell, Charles. The Opioid Toll, by Neighbourhood, that Alberta Tried to Hide. *The Tyee,* 6 July 2022, https://thetyee.ca/News/2022/07/06/Alberta-Opioid-Toll-By-Neighbourhood/.

Scherer, Jay, David Mills, and Linda Sloan McCulloch. *Power Play: Professional Sport and the Politics of Urban Development.* U of Alberta P, 2019.

Scherer, Jay, Judy Davidson, Rylan Kafara, and Jordan Koch. "Negotiating the New Urban Sporting Territory: Policing, Settler Colonialism, and Edmonton's Ice District." *Sociology of Sport Journal,* Vol. 30, No. 2, 2021, pp. 111–19, DOI: 10.1123/ssj.2020-0113.

Scherer, Jay, Rylan Kafara, and Jordan Koch. "The Spectacle of Settler Colonial Urbanism, Racialized Policing, and Indigenous Refusal of White Possessive Logics." *Settler Colonial Studies*, Vol. 13, No. 3, 2023, pp. 1–22, DOI: 10.1080/2201473X.2023.2195044.

Scherer, Jay, Rylan Kafara, and Judy Davidson. "A Few Weeks in Dirt City: Sport-Related Gentrification, Mobilizing Resistance, and the Art of Failure in Edmonton, Alberta." *Journal of Sport and Social Issues,* Vol. 46, No. 6, 2022, pp. 499–523, DOI: 10.1177/01937235221094063.

Scott, Conrad. "'Changing Landscapes': Ecocritical Dystopianism in Contemporary Indigenous SF Literature." *Transmotion*, Vol. 8, No. 1, 2022, pp. 10–38, DOI: https://journals.kent.ac.uk/index.php/transmotion/article/view/979.

Simpson, Audra. *Mohawk Interruptus: Political Life across the Borders of Settler States.* Duke U P, 2014.

Smith, Neil. *The New Urban Frontier: Gentrification and the Revanchist City*. Routledge, 1996.

Spencer, Donna. "NHL and Its Labs Say COVID-19 Testing Doesn't Overlap with Public's." *CTV News*, 6 August 2020, https://edmonton.ctvnews.ca/nhl-and-its-labs-say-covid-19-testing-doesn-t-overlap-with-public-s-1.5053254.

Sportsnet. "Jets fly over Rogers Place during National Anthem." *Sportsnet*, 25 September 2020, https://www.sportsnet.ca/nhl/video/jets-fly-rogers-place-national-anthem/.

Sousa, Aaron. "'We Need Action': Treaty 6 Chiefs Declare Opioid Crisis Emergency." *CBC News*, 10 July 2023, https://www.cbc.ca/news/canada/edmonton/treaty-6-addictions-opioid-crisis-state-of-emergency-1.6902573.

The Progress Report. "Exclusive: Alberta Government Quietly Gave the NHL $4 Million during Pandemic." *The Progress Report*, 29 November 2020, https://www.theprogressreport.ca/exclusive_alberta_government_quietly_gave_the_nhl_4_million_during_a_pandemic.

Thomson, Euan. "Who's Really behind Alberta Drug Policy?" *Drug Data Decoded,* 27 April 2023, https://www.drugdatadecoded.ca/p/whos-really-behind-alberta-drug-policy.

Thomson, Euan, Michelle Robinson, Terrill Tailfeathers, and Will Mawer-Cardinal. "Clearing the Plains 2.0: Alberta Drug Policy Is Indigenous Genocide." *Drug Data Decoded,* 16 May 2023, https://www.drugdatadecoded.ca/p/clearing-the-plains-20-alberta-drug.

Vizenor, Gerald. "Aesthetics of Survivance: Literary Theory and Practice." *Survivance: Narratives of Native Presence*, edited by Vizenor, U of Nebraska P, 2008, pp. 1–25.

Vizenor, Gerald. *Fugitive Poses: Native American Indian Scenes of Absence and Presence.* U of Nebraska P, 1998.

Wolfe, Patrick. "Settler Colonialism and the Elimination of the Native." *Journal of Genocide Research,* Vol. 8, No. 4, 2006, pp. 387–409. DOI: 10.1080/14623520601056240.

9 Affiliation as Environmental Justice in Three Climate Novels

Claire P. Curtis

Unsurprisingly, climate fiction has become a prominent genre in contemporary literature. Both speculative (e.g., Margaret Atwood's MaddAddam series) and realist novels (e.g., Jenny Offill's *Weather*) now consider the impact of climate change and its myriad crises and consequences. Within both types of climate fiction, the speculative and the realist, questions about justice emerge. How might changing climate conditions impact the ways humans live together, and live together with nonhumans? In climate fiction, characters face a variety of challenges to living together justly: extreme weather, displaced peoples, political unrest, resource scarcity, loss, and disease. This chapter focuses on the particular subgenre of climate fiction that depicts the impacts of pandemic diseases. The chapter considers the role of pandemics in our understanding of environmental justice through a reading of three novels. Drawing from Martha Nussbaum's capabilities approach, specifically the capability of affiliation, the following illustrates the ways in which pandemics intersect with concerns about justice.

Leigh Richards' *Califia's Daughters* (2004), Emily St. John Mandel's *Station Eleven* (2014), and James Bradley's *Clade* (2015) all include pandemics of different kinds and lethality. The characters' responses to those pandemics, in turn, produce different understandings of both the content and the extent of justice in the world. In its description of a just post-pandemic community, *Califia's Daughters* concentrates on the relationship between humans and the nonhuman world. *Station Eleven*, on the other hand, focuses entirely on how humans can survive in groups after the devastating Georgia Flu. Both novels draw sharp lines between pre- and post-pandemic forms of life. *Clade*, however, resists both of these binary oppositions in favor of a complicated interplay between humans and the world in which they live. Bradley's novel represents a new kind of climate fiction, where the interlocking impacts begin in the present and there is no catharsis from a catastrophe ending or being solved. Instead, *Clade* sets its storyline in an in-between space where environmental justice is at stake. Negotiating that space, as both a reader and a citizen of a rapidly changing world, is a crucial task.

This chapter proceeds in three stages. I first outline Martha Nussbaum's theory of justice as focused on capability, what "we are able to do and be" (70).[1] Nussbaum's capability of "affiliation" highlights how our interactions with one another and, I argue, with the wider world is central to living justly. Next, I explain how

DOI: 10.4324/9781003345770-13

pandemics raise pressing issues about the capability of affiliation in the context of environmental justice. Pandemics are particularly challenging to human interaction, and so these novels all offer visions of living justly together despite that challenge. The final section uses each of the novels to consider the interrelation between living together and living within the natural world. Once again, I argue that *Clade* understands this interrelationship in the most complex and practical way.

Martha Nussbaum, Capabilities, and Environmental Justice

Nussbaum draws a contrast between traditional distributive theories of justice and what she calls the capabilities approach. Distributive theories, by their very name, consider the justice of the distribution of a given set of resources. In the environmental context, distributive justice would refer both to the potentially disparate impact of certain harms, such as pollution, toxic waste, or food instability, and the potentially disparate access to certain benefits, such as clean water, clean air, or access to green space. The capabilities approach, by contrast, focuses not on resources but on individuals, asking, "What am I able to do and be?" (Nussbaum 70). From an environmental perspective, the capabilities approach expands our understanding of justice, by asking how we might cope under new and different conditions.

Nussbaum posits a list of ten capabilities or central ways of doing and being that she claims are necessary for a flourishing life. These ten capabilities sketch a minimum threshold for justice. The complete list is life; bodily health; bodily integrity; senses, imagination, and thought; practical reason; emotions; affiliation; play; other species; and control over the political and material environment (Nussbaum 76–77). Here I focus on affiliation as a particularly useful capability for understanding environmental justice in these (and other) climate novels. Affiliation is defined as "living with and toward others" (Nussbaum 77). By including the word "towards," Nussbaum is stressing that we should act in ways that not only acknowledge but also positively encourage, the flourishing of others (whether they be human or not). Justice requires that we see others not simply as potential means to our own ends but as ends in themselves. These others with their own ends must be both protected and encouraged.

Affiliation also includes having the "social bases of self-respect and non-humiliation" (Nussbaum 77). The social bases might include guaranteed schooling for all or a developed system of anti-discrimination law. From the perspective of environmental justice those social bases include how the changing conditions of the world impact the ways in which we live together. As David Schlosberg notes, we live in the world and "[t]he discussion of climate change illustrates the centrality of this connection between the condition of the natural world and the material experience of everyday life" (48). Thinking about affiliation alongside environmental justice means recognizing how our personal interactions happen within changing environments. The particular scenarios presented in climate fiction can help us understand "*how* environmental resources and conditions contribute to people's capabilities to live physically healthy lives of opportunity and meaning, and *who* is benefitting from policies that succeed or fail in protecting those resources and conditions" (Holland 201).

Holland argues that using a capabilities approach to environmental justice shifts attention away from resource management (as in the distributive approach) to the values that underscore living together (97). We can understand the moral impact of climate change by examining its specific effects on our capacity for affiliation.

Pandemics

Pandemics pose a special and direct challenge to affiliation. As a mode of contagion, a pandemic elicits fear of both strangers and loved ones. Fear encourages isolation: the best response to a pandemic is for individuals to avoid others. Unlike other cataclysmic events, pandemics are not set up for a truly collective response. Here, pulling together means staying apart. Because pandemics challenge what humans are able to do and be, post-pandemic imaginaries are able to raise a fundamental question about how we live with one another, and particularly about how to repair our loss of connection to other humans, to nonhumans, and to the environment in which we all live.

Pandemics are also central to the potential risks of continued climate disruption. The very same conditions (heat, displacement, habitat loss) that might themselves alter the ways in which humans interact with one another are also the conditions conducive to disease emergence and spread. This means that the potential fear of the other is not necessarily limited to a particular disease at a particular time, since a future (or more virulent) disease is always possible. We now live under the *expectation* of that future.[2] In this sense, the fear that pandemics produce is not limited to the specific effects of the pandemics themselves. Rather, pandemics prompt a generalized anxiety about whether and how we will be able to live together, about our very capacity for affiliation.

Novels

Fiction provides an experimental space that is otherwise not available to political theorists. We cannot ask groups of people to live under manufactured sets of conditions and then evaluate the success or failure of their living together. But novels provide just such a space. The three novels I have chosen to analyze here, Richards' *Califia's Daughters*,[3] Mandel's *Station Eleven*, and Bradley's *Clade*, all include pandemics in their speculative futures and each novel follows a cohort of characters through a kind of aftermath. *Clade* and *Califia's Daughters* are primarily works of climate fiction (although *Califia's Daughters* was published in 2004, prior to the popular use of that term). *Station Eleven*, framed around its deadly global pandemic, may not be a classic climate novel, but it is fully a novel of the Anthropocene.

In *Anthropocene Fictions* (2015), Adam Trexler observes that the Anthropocene is marked in part by a recognition that "later disasters are contiguous with our moment" (4). In each of these novels, the pandemics and other effects of climate change are extensions of our current situation. These novels are not concerned with potential mitigations to climate change, as in the climate fiction of Kim Stanley

Robinson (especially his 2020 *Ministry for the Future*). Instead, these novels imagine human interactions under changed conditions. In his reading of *Station Eleven*, Pieter Vermeulen argues that the novel uses objects, for example, the comic book and items from the character Clark's museum, to provide a "tenuous sense of continuity" (19) for the reader. I argue here that the human connections in the novel also provide this continuity. Briohny Doyle makes a similar point about *Clade*, suggesting that human interaction "underscores the creativity, alliances and relationships of care that enable life in climate crisis" (169). It is these novels' emphasis on human interaction that allows for a reading that focuses on the possibility of justice through the idea of affiliation in the context of climate crisis.

I begin with brief overviews of these three novels. *Califia's Daughters* includes a past plague resulting from a time period of terrorism, "virus and nonphage, genetic modification and dirty bombs" (30). The novel is set in a near future after a series of economic, political, and environmental crises. The primary characters live in a small community in a valley outside of what was once San Francisco. The pandemic has been over for many years, leaving behind a deeply skewed sex ratio where few men live to adulthood. The community is thriving, but the plot of the novel takes Dian, the main character, outside her community to investigate a community further north as a possible ally against an encroaching threat, the tyrannical and power hungry "Queen Bess" from Oregon. Her journey eventually leads her to Ashtown, an ally of Bess. My analysis focuses on the contrast between the productive tranquility of the valley community and the very different form of community represented by Ashtown.

Station Eleven is set both before and after a deadly flu. The novel follows a set of characters loosely connected to Arthur Leander, an actor who dies of a heart attack on the date of the arrival of the flu in Toronto in the opening chapter. The analysis here focuses on the sections of the novel after the Georgia Flu, and particularly on the community of the Traveling Symphony, a group of Shakespeare performers and musicians that includes Kirsten, a woman who, as a child, acted alongside Arthur in his final performance. The Symphony travels a route through what was Northern Michigan, performing for scattered groups of survivors. In doing so, they encounter a self-titled prophet (Arthur's son) forcibly recruiting followers and eventually join up with a survivor community led by Clark, who was Arthur's best friend, in the former (fictional) Severn City airport.

Clade opens with Adam on a scientific expedition in the Antarctic, awaiting a phone call from his wife concerning the success of their most recent round of fertility treatments. The novel then moves forward from Adam in this slice of time to his death some 60 years later. Each chapter brings a different cohort of people together, but all are somehow connected to Adam. The order of the chapters follows a loose genealogy: Adam, Ellie, their daughter Summer, her son Noah, Noah's friend Lijuan, Lijuan's eventual husband Dylan, and their daughter Izzie. The characters' disparate stories are also stories of economic decline, civil unrest, climate refugees, the demise of bees, a pandemic, and the discovery of an alien signal. While this might sound like a confusing series of events, the novel instead reveals a web of connection between a set of people embedded in a world that is changing and the adjustments these changes require them to make. *Clade* does not rely on external threat to cement alliances; nor does the narrative use any singular catastrophe to motivate change. Instead, it highlights living "with and towards" one another in a changed and changing world.

The following examination considers how these three novels depict their characters' capacities for affiliation. I first consider the fact of affiliation: in what ways and to what extent do these novels show their characters living with and toward one another? How important is affiliation to the characters, and what are its boundaries? I then consider the grounding of affiliation: to what extent do these characters live in communities that understand both the importance of affiliation and seek to nourish it? What can the presence and foundational importance of affiliation show us about the presence and the content of justice in these communities?

Finding Affiliation

Affiliation is challenged both by disease and by isolation as a response to disease. Disease and isolation exacerbate vulnerability—the bodily vulnerability of sickness and the emotional vulnerability of isolation. Vulnerability is the precondition for "living with and toward others" (Nussbaum 77). All three novels share the sentiment expressed succinctly by Adam in *Clade*: "I just want both of you to be safe" (Bradley 209). Pandemics undermine safety. But what do we even mean by safety? Complete safety is illusory. The choice of isolation by characters in these novels highlights this illusion. Jeevan and Frank in *Station Eleven* are sheltering in Frank's Toronto high-rise apartment. Lijuan, in *Clade*, leaves the city of Sydney with Adam and Noah to wait out the pandemic in Adam's family cabin. The people of the valley community in *Califia's Daughters* retreat during their time of crisis. Isolation has largely solved the fear of infection. But isolation is not the solution to how to live. Nor is finding a purported and isolated safe harbor the solution to climate disruptions. Affiliation is both the capability needed to respond to these mounting crises and one sign of a potentially just community when in the face of crisis. While many postapocalyptic novels revel in details and minutiae of addressing the impact of these crises (e.g., the long descriptions of food production and the creation of a laundry in James Howard Kunstler's *World Made by Hand* (2008), or the description of salt harvesting in Pat Frank's *Alas, Babylon* (1959)), these three novels are far more focused on the human interactions that are impacted by the crisis at hand. I explore this focus on connection as one way of seeing the importance of affiliation as a feature of justice.

Califia's Daughters describes the people of the valley as flourishing physically and emotionally. This flourishing is no accident, but the result of a purposeful system of living together. While the novel opens with the arrival of a potential threat from an unidentified group of people approaching their home space, the description of typical life that follows is one of harmony:

> the workday noise of an active community—children shouting and adult voices raised in work and song, the rumble and thump of the mill machinery,

> the echo of the hammer and the rasp of saw and the jingle of harnessed horses pulling plow or cart. The myriad sounds that made up the daily voice of the Valley. (Richards 7–8)

This description focuses on the people, the animals, the interactions, the ways in which they live with, work with, and as becomes clear in the novel, respect one another. This is a community whose collective work signals its collective endeavor, to move beyond survival to flourishing.

Califia's Daughters describes the Valley community as working with and toward one another and the natural world. Community members are described as training and caring for dogs and horses of the community and nurturing the crops and trees, streams, and ponds. Dian, head of security, came to the community as an infant found abandoned at the side of a road, with an enormous dog as companion. Her dogs, and even the horses of the community, feature as characters in the novel as much as the humans. This community still experiences fragility, attempting to survive on its own among both potentially hostile travelers and the lasting consequence of bio-terrorism, leaving lowered birth rates and higher mortality rates among boys. When Judith, Dian's sister, gives birth to a boy, she is warned by another woman in the community that, if she cannot wholly love that child out of fear of his early death, then she has to give the child to someone who will. That woman notes that "we owe it to him, as a small human being, to make what days he has as full and comfortable and as filled with love as we possibly can. If he lives, that love will make him a better person. If he dies, it will come near to killing us" (159). This baby boy is accorded the dignity of being cared for with love despite the likelihood of his shortened life. He is someone in the community, and if Judith cannot live "with and toward" him then another member of the community will. This sentiment is echoed throughout the novel: caring for others is the backbone of building community, whether those others are people or dogs or the world itself.

People and nonhuman others are respected as they are, and the community thrives as a result. For example, Isaac's son Teddy, who has not spoken since his mother's death, is accommodated and his particular interest in Dian's dogs is nurtured. Children are not asked to change their ways to make the life of the community easier for the adults. The community sees value in each member. Community members may not always be in agreement, but even their disagreements are worked through with a recognition of the dignity of each.

While *Station Eleven* is also a novel about human connection, it is not a novel about connection to the wider world of plants, other animals or the landscape. The Traveling Symphony is a close-knit community with little connection to the larger world. *Station Eleven* is set largely in the detritus of the built, human world. Humans have retreated to the vestiges of the human-made: former airports, fast food restaurants, and box stores. It is only in art and human connection that affiliation flourishes. But this makes for a novel whose natural world is merely a backdrop. Nature is not a source of human meaning or flourishing. There is little to celebrate in the larger environment. The characters note the heat, the inhospitable forest, the crumbling buildings, and the occasional "ferals" (humans no longer living in the community). The external world is seen only as a source of discomfort and threat. Affiliation here is limited to small-scale human interactions. The Traveling Symphony (and the community living in the Severn City airport) recognizes the importance of affiliation as a principle of justice, but affiliation here has limits. Humans in *Station Eleven* do not belong to a wider world; they merely move through it.

The Traveling Symphony is a collection of people who struggle with and yet continually commit to one another. The assertion that "the problem of the Traveling Symphony was the same problem suffered by every group of people everywhere since before the collapse" (46) is followed by a full page detailing who resents whom, along with the bits and pieces of negative human interactions. At the end of this list comes a general assertion of the value of community itself:

> This collection of petty jealousies, neuroses, undiagnosed PTSD cases, and simmering resentments lived together, traveled together, performed together 365 days a year, permanent company, permanent tour. But what made it bearable were the friendships, of course, the camaraderie and the music and the Shakespeare, the moments of transcendent beauty and joy when it didn't matter who'd used the last of the rosin on their bow or who anyone had slept with. (Mandel 47–48)

This passage affirms not only the role of art as a means of tying people together but also the value of being tied together in human flourishing. There is little about this description that is particular to a group of people living post-pandemic; the struggle to live fruitfully with others is not specific to the postapocalyptic landscape. But that landscape heightens, for the reader and for the characters, the fact and the stakes of living together.

In contrast to the two novels discussed above, *Clade* rejects both *Califia's Daughters'* vision of a kind of harmony with nature and Station Eleven's depiction of humans living uncomfortably adjacent to nature. From the start of the novel, natural forces are in play in ways that cannot be controlled and for which humans are largely responsible. The pandemic is simply one among many crises of the novel: a devastating flood in England, the death of the bees, the arrival of climate refugees, and rising seas and temperatures. But the characters of the novel keep going on, not by defeating some threatening outside force, but by continuing to move forward. This is not a novel of clever mitigations, but it is a novel that provides hope in a climate changed world. The possibilities of justice for humans, for nonhumans, and for the earth itself require a willingness to persist.

In *Clade*, the interpersonal is seen through the arc of Adam's family story, which is a story of loss and connection. The pandemic, as with the other events that shape the characters, does not cause a break, a moment after which people must learn to live differently. This reveals possibilities in the ways in which we look for justice in climate fiction. We need not read for apocalyptic events and a clean slate from which to start over. Instead, we can follow characters who adjust, survive, and move ahead while still maintaining connections.

Pandemics emerge among both humans and bees. The demise of the bees comes first, as described by Adam's ex-wife, in an art project. Ellie's interest in bees happens alongside her meeting and friendship with Amir, a climate refugee who has been caring for the few wild hives found near Ellie's home. The interconnections among the bees, the drowning of Bangladesh, the death of Amir's wife and child and his experience in camps in Australia, and a visit by Noah (who goes missing) all come together through Ellie. Ellie muses about the long human connection with bees: "the idea that humans have shared the world with these creatures for so long fill[ed] her with something that is not quite wonder, not quite grief, but somehow both" (Bradley 147). The bees, we learn in the next chapter, die. But the novel is not directed toward this or any of its potential crises. Excepting Adam's experience in the flood, these crises are not seen first-hand. *Clade* is a novel about a series of catastrophic events, any of one of which might be the focus of survival and transformation in another novel. Here, instead, these events are simply the backdrop to a slowly changing way of living.

We initially learn of the human pandemic, a COVID-19-like viral illness that begins in China, through Lijuan's experience of retreat with Noah, as detailed in her journal. The entries recount their living in a cabin on the edges of a forest, getting supplies from nearby towns whose stores have been looted, and Lijuan's increasing worry about her mother, who has gone back to China to care for her sister. Alongside these pandemic-related events is a journal entry titled "these are things we've lost," organized as a list: "Birds, Bananas, Tigers, Frogs, Bees, Coffee, Polar Bears, Coral" (Bradley 207). Following this is another list of what has been saved: "Seeds, Elephants, Dolphins, Each other" (Bradley 207). This entry is never explained. It follows a walk in the forest where Lijuan calls up on her Google Glass-style lenses a simulation of birds from the past. She shares the simulation with Noah, saying, "I've been thinking about what it must have been like out here before the change began, what the forest was like when there were still birds" (Bradley 206). Lijuan lives in a world that seems catastrophically different from ours. But her journal and its descriptions of evacuation with Adam and Noah are not a catalogue of how she survived. Instead, we read a fairly mundane outline of her experience in this remote cabin. She is worried about her mother and somewhat uncomfortable with Adam and Noah, neither of whom she knows very well. She wanders in the area around the cabin, thinking about what used to be. It is not always clear how species loss or the drowning of cities matters to her, but this may be part of the point. The particularity of Lijuan's experience forces us to think about the more general problem of living in a radically changed world in a different way.

The only framework for this more general question in *Clade* is in its continued return to the collection of people around Adam. In an interview about the novel, Bradley notes that the title, *Clade*, describes a

> scientific term for a group of organisms with a common ancestor and comes from the Greek word *klados*, or branch. I chose it because it describes the structure of the book, but also because I love the way the word itself is simultaneously so sleek and futuristic, but also echoes 'glade,' so has all those

> associations of sacredness and beauty it has. I hope the book has some of that same combination of qualities. ("Interview")

The "clade" of the title captures not simply the fact of this familial set of connections, but also a way of seeing the world through the idea of connection.

Grounding Affiliation

Detailing the mere fact of affiliation is not sufficient for understanding the importance of this capability for the potential of justice of these communities. Instead, we must go on to consider the ways in which the novels recognize how affiliation (or any other capability) is nurtured or can be undermined in these communities. Two of the novels recognize the importance of affiliation as something to be nurtured by presenting contrasting groups of people who reject affiliation. The Prophet and his followers in *Station Eleven* and the leaders of Ashtown in *Califia's Daughters* threaten the primary communities of the novels with violence.

These violent groups share certain characteristics: a commitment to a strict hierarchy, a rejection of any dissent or resistance to authority, and a clear identification of who does and does not belong to the community. In *Califia's Daughters*, Dian joins the Ashtown Guards in an attempt to free Robin, the man who had nursed her back to health after she was attacked on the road. She believes that she can join the guards in order to find and rescue Robin, bringing him back to the Valley. Breaker, the Captain of these guards, tells Dian,

> discipline here is putting up with anything—*anything*—that I say you put up with. I say you crawl, your chin is on the ground. I say you submit to a strip search, you spread 'em before I finish the sentence. That's what discipline means in my guard, and frankly, I don't think you can cut it. (Richards 367)

Dian assumes that she can hold out—maintain emotional distance—in order to fulfill her goal of freeing Robin.

Breaker, the Captain of the Ashtown Angels, is clearly not interested in Dian as a person. Dian is merely a tool expected to obey whatever the Captain might command. Breaker notes at the end of Dian's initiation, "you are mine" (Richards 389). These are not conditions under which affiliation can be nourished or even maintained. Guards treat lower-ranked guards or others in Ashtown with disrespect and the expectation of immediate obedience. Breaker tempts Tomas, Dian's dog, away from Dian, claiming that no other guard may have a potential weapon that she, as Captain, cannot use. This is not only harmful to Dian, but it also shows a lack of respect for Tomas. The depiction of Ashtown is not ethically difficult to evaluate. Ashtown provides a clear contrast to the community in the Valley, where security is also an issue, but where the protection of the community does not mean the subordination of all under the control of Dian as head of security. Dian may lead security, but she does so as one who works with her team of people and dogs as equals in dignity.

The Prophet in *Station Eleven* shares with Breaker an ideology of total adherence. He persuades his followers by claiming to be leading them as a "chosen" set of people who survived the Georgia Flu for a reason—"that flu was our flood" (Mandel 60). But even after the flu, there is still "culling" to be done, and those who follow the Prophet are the "light" (Mandel 60) and they "are the pure" (Mandel 61). The Prophet maintains a harem of child brides and has arrived in St. Deborah by the Water after traveling south, killing and amassing weapons. A system that divides the populace into the damned and the saved clearly undermines the ability to live with and toward others. The depiction of the Prophet highlights the particular choices made by the Traveling Symphony and the people of the Severn City airport to come together in recognition of affiliation as an essential part of living a dignified life.

The similarities in the examples of Breaker and the Prophet are telling. The authority of the one in charge is largely specific to that one person. These are both individuals who have amassed power on the basis of their own (alleged) unique qualities. The figures of power project themselves as distinct from anyone else, and so neither they nor the others around them can interact with that person as full human beings. Second, the power that each holds is backed by a system of fear and violence. There is no trust in the value of their leadership. The followers mistrust those in charge. The leaders both mistrust their followers and fear the repercussions of resistance. Finally, in each case, there is a strict differentiation between those who follow and those others who do not. For the Prophet, this is couched in a religious language of who has "seen the light" and a message of cleansing the earth. For Breaker, the differentiation is simply about power—who has it and who does not.

Clade, by contrast, offers a more subtle, ambiguous, and perhaps more realistic consideration of the structures necessary for affiliation. Instead of focusing on a potential enemy who is actively violating the dignity for others, this novel offers a chance for the reader to discover for themself the conditions needed to sustain affiliation—or the conditions that undermine it. For example, when Ellie meets Amir, the beekeeper, she comes face-to-face with the refugee situation she had previously ignored as "a constant rumble of anger and paranoia" (Bradley 167). After meeting Amir, she reads about the refugee camps, with "the random harassment by police, the detention and forced expulsion of anyone the government deems undesirable" (Bradley 167). This description, not of a constructed external enemy but of the actions of her own government, threatens to undermine affiliation in the same ways as the actions of Breaker or the Prophet. Refugees are being singled out, harassed, and denied services and dignity.

Amir explains to Ellie what it is like to live under these conditions, noting,

> we don't just need access to hospitals, we need medicine, schools, jobs, not to be frightened all the time. To be able to buy food without being terrified somebody will become suspicious because we're using cash, or report us to the police because they don't recognize us. (Bradley 174)

Amir is noting a need for recognition. He wants to be seen as one of the community, and not to be feared as a source of suspicion. He is advocating for a capabilitarian approach to justice, suggesting not a list of resources needed but rather a set of conditions necessary to live a life of dignity.

Readers may well dismiss Ellie's own comfortable ignorance about the situation of people like Amir. She knows why these climate refugees are in Australia, noting to herself that "Bangladesh is gone as is much of Burma and coastal India" (167). Where did she think that people would go? Surely, she should have been paying more attention. I do not think Bradley is presenting Ellie's ignorance as particularly unusual; he is not justifying it, but he is poking the reader about their own ignorance of such situations. *Clade* illustrates a point that readers of speculative fiction would do well to remember: we do not need to create some extravagant tyrant to see and consider the lives of people whose dignity has been ignored. Fictional accounts of systemic humiliation do not need the more extreme humiliations of so-called Prophets or "breakers." *Clade* quietly illustrates the ongoing humiliation of those who suffer through political violence, economic downturn, or environmental destruction, whose suffering is made worse by being treated as disposable.

There is no grand solution offered to state violence against refugees. Bradley's novel is not one of solutions. Crises like catastrophic floods and pandemics divide people. But how one responds to crisis illustrates the extent to which one values justice and affiliation in particular. In *Clade*, we never learn if policy change has happened, although we can imagine that, after the pandemic (which occurs in the following chapter), the presence of refugees matters less. The two clues we do receive come in very different ways. The first is the presence of Amir with Ellie in the final chapter of the novel, which signals his continued presence in her life. He was not introduced as a character simply to teach Ellie about refugees. The second is the inclusion of the alien signal.

The last section of the novel describes Noah, Adam's grandson, identifying a signal from space. That signal, initially difficult to decipher because of incorrect expectations of how such a signal would manifest, is eventually discovered to be in alien words: "whoever they are, whatever they are, they have not chosen the language of numbers or mathematics, they have chosen words, and not in order to be understood, but merely to speak" (Bradley 283). How do we read this inclusion of aliens in *Clade*? It matters that this signal is not the beginning of a crisis. But it does encourage a focus on the future. Noah notes that

> it will be five hundred years before the message reaches SKA-2165, another five hundred before anyone hears back. He will be gone by then, as will Adam, Ellie, Lijuan, Dylan, Jin, Amir, all of them vanished into the distant past, their passage through the world remembered, if at all, by a handful of video recordings, a scattering of data traces. (Bradley 285)

What and who will be there to receive this message in 1,000 years? Noah does not know, and neither does the reader. But raising this question in this novel of interconnection primes a reader to think about possibilities. What might life be like in 500 years? Do we imagine humans present to receive this signal? What will that world in be like?

By raising these kinds of questions, climate fiction forces us to think about justice, whether that means justice for humans, for nonhumans, or for the larger world. Because climate change represents such an overwhelming threat, it might seem easiest to think through these questions in postapocalyptic settings that are stark and limited, like the seemingly utopian space of *Califia's Daughters* and the denuded, human-only space of *Station Eleven.* But by presenting a world not so different than the one where we are now, *Clade* allows us to think through just responses to climate change in a way that is ultimately more practical. When we think about our affiliations with others, we do not imagine ourselves in artificial scenarios of small, isolated communities confronting existential threats. Nor does thinking effectively about climate change imply imagining ourselves in such scenarios. Confronting climate change does not mean waiting for catastrophe; nor does it mean expecting that, when catastrophes do occur, the means of response will magically emerge. Climate fiction reminds us of what we have in our world, of what we have not yet lost, and why we should work to save it.

Notes

1 I reference Nussbaum's *Frontiers of Justice* here as one of the first full-length discussions of the capabilities approach. Her later works utilize the same ten capabilities and descriptions.

2 This expectation is illustrated in Mandel's 2022 novel *Sea of Tranquility*, which includes a character who is an author that has written a very popular pandemic novel, which includes the line, "we knew it was coming" (Mandel 86–87). That author then meets someone at a book reading who has tattooed the phrase on her arm (67). The doubling and tripling of meaning and experience here illustrate a piece of our new condition of living.

3 "Leigh Richards" is a pseudonym for the mystery writer Laurie King.

Works Cited

Bradley, James. *Clade*. Titan Books, 2017.

Doyle, Briohny. "Hope out of Stock: Critical and Melancholic Hope in Climate Fiction." *Dystopian Emotions: Emotional Landscapes and Dark Futures*, edited by Jordan McKenzie and Roger Patulny, 1st ed., Bristol UP, 2022, pp. 159–175. *JSTOR*, https://doi.org/10.2307/j.ctv24cnssh.14. Accessed 8 June 2023.

Frank, Pat. *Alas, Babylon*. 1st Harper Perennial Modern Classics ed. Harper Perennial Modern Classics, 2005.

Holland, Breena. *Allocating the Earth: A Distributional Framework for Protecting Capabilities in Law and Policy*. Oxford UP, 2014.

"Interview with James Bradley, Author of Clade." The Qwillery, 5 September 2017, https://www.theqwillery.com/2017/09/interview-with-james-bradley-author-of.html. Accessed 9 January 2023.

Kunstler, James Howard. *World Made by Hand.* Grove Press, 2008.

Mandel, Emily St. John. *Station Eleven*. Vintage Books, 2014.

———. *The Sea of Tranquility*. Knopf, 2022.

Nussbaum, Martha. *Frontiers of Justice*. Harvard UP, 2006.

Richards, Leigh. *Califia's Daughters*. Bantam Dell, 2004.

Schlosberg, David. "Theorising Environmental Justice: Expanding the Discourse." *Environmental Politics*, vol. 22, no. 1, 2013, pp. 37–55, https://doi.org/10.1080/09644016.2013.755387.

Trexler, Adam. *Anthropocene Fictions: The Novel in a Time of Climate Change*. U of Virginia P, 2015.

Vermeulen, Pieter. "Beauty That Must Die: *Station Eleven*, Climate Change Fiction, and the Life of Form." *Studies in the Novel*, vol. 50, no. 1, 2018, pp. 9–25. *Project MUSE*, https://doi.org/10.1353/sdn.2018.0001.

10 "A vortex of summons and repulsion"

The Productive Abject, Posthumanisms, and the Weird in Charles Burns' *Black Hole*

Benjamin Horn, Jayde Martin, and Alice Seville

Following the outbreak of COVID-19, the idea of human separateness from our natural environment is increasingly difficult to maintain. Like other zoonotic diseases, COVID-19 eroded any symbolic "delusive envelope" between humans and other animals, and our activity amid a wider biosphere. The breakdown of an artificial separation between human and nonhuman (or between a delusive "outside" environment and a treasured "inside" of networked, cosseted human society) became an inescapable reality in a post-COVID-19 world. Epidemiologist Cristina O'Callaghan argues that increased human contact with other biological communities can catalyze the spread of diseases[1] between these communities, writing that "consumption of fossil fuels for energy, deforestation and the conversion of natural habitats into farmland or extensive livestock" not only produce significant greenhouse gas emissions, but "facilitate the emergence of new zoonosis, such as SARS-CoV-2 [COVID-19]" (O'Callaghan-Gordo and Anto). In short, diseases can be treated as an uncomfortable synecdoche for material embeddedness. Through their interpenetration of treasured boundaries, zoonotic diseases force a wider confrontation with humanity's ecological embeddedness. We must all face such embeddedness in an era of late capitalism to affect a necessary breakdown of the self/other distinction. Awareness of ecological embeddedness also has the power to undermine the ideological derivations of this distinction. As we will explicate, these derivations enable biodiversity loss, pollution, and global warming.

Furthermore, drawing on the work of Mark Fisher, COVID-19 exists on the register of the "weird" in its post-dualistic breakdown of artificial boundaries. The virus possesses a form of "real externality" (Fisher 16), a sense that it "does not belong" (Fisher 13), plus the feeling that it "should not exist" (Fisher 15), while revealing the inadequacies of our categories for fully making sense of the world (Fisher 15). Encounters with real externality, or "the outside" (Fisher 16) corrode conceptual and bodily boundaries, effectuating a revisionary critique of hegemonic forms of subjectivity:

> The encounter with the outside often ends in breakdown and psychosis [...] a catastrophic integration of the outside into an interior that is retrospectively revealed to be a delusive envelope, a sham. (Fisher 16)

DOI: 10.4324/9781003345770-14

Building on the lived experiences of this breakdown of boundaries, we will examine forces of corrosion, transformation, and liminal "porousness" (Pizzino 159) in the treatment of virality and ecological embeddedness in Charles Burns' graphic novel, *Black Hole* (2005).[2] We posit that Burns' breakdown of boundaries is nominatively posthumanist and ecocritical, highlighting how the boundaries between self and other are mutually permeable. We analyze Burns' aesthetic mode as one of defamiliarization, purposefully elucidating the graphic novel's manifest themes of empathetic ecology, sociality, and mutualism.

Our focus is upon the Jonathan Cape 2005 edition of Burns' comic collection (1995–2005), which draws together, in an interconnected narrative of stunning "contiguity" (Kunyosying 564), a decade of serious work, rendering in "dichromatic" (Protic and Finlayson) panels the troubled lives of Keith, Todd, Chris, Rob, Eliza, Dave, and Rick, among others. Burns' *Black Hole* follows the lives of these Seattle teenagers as they are affected by, or become infected with, the Bug, a fictional virus. The Bug is a sexually transmitted infection that causes mutations, ranging from animal-like alterations to horror-type aberrations. We elucidate how Burns details, through his aesthetic registers and recursive visual signifiers, a state of social claustrophobia at the outset of his narrative set within the teens' disordered Seattle community. This state of claustrophobia is then interpenetrated with the effects and social consequences of the Bug: an abjecting yet also illuminating force in Burns' narrative, since it draws stark and powerful attention to the teens' status as ecologically embedded creatures.

Through his art style, Burns renders ecological embeddedness conspicuously visible, forcing his characters to realize that the dualistic binary of inside-outside is "delusive" (Fisher 16). By conspicuous visibility, we refer to Burns' contiguous flows of objects, which elide the foreground/background distinction typical of comics. Doing so formally expresses the inescapable continuity of characters with their environs. Burns uses what we term simultaneous objects to signify this interconnectivity. Simultaneous objects are objects that present as themselves but can also be interpreted as something else. For example, Chris is a major character who suffers from low self-esteem, plus alienation from the patriarchal facets of her close-knit community. Chris' skin frequently appears as if it were foliage, parchment paper, a burning joint, a disembodied nervous system or a removable, sheath-like substance like that of a snake. By using simultaneous objects, Burns blurs the boundary of inside-outside in a way that is unmistakable: the visual metonymy of Burns' simultaneous objects elucidates the 'sham' (Fisher 16) of symbolic orders, as inside-outside boundaries are corroded and thus consequently so is a network of distinct symbolic categories.

Our use of the concept of "symbolic orders" here refers to the world of communication, language, interpersonal relations, and ideological conventions. Symbolic orders mediate human experience of factual reality without being reducible to individual psychologies, although they interpenetrate with them. Symbolic orders shape how things "make sense," as well as our taken-for-granted assumptions. Žižek notes that symbolic orders supplement the "relationship of 'external' factual reality and 'internal' subjective experience" (Žižek 13). Symbolic orders and their taken-for-granted assumptions are achieved through "identification" (Žižek 116) of the subject with a signifier, by which it achieves "self-identity" (Žižek 116). This process involves alienation, as the subject identifies with an ideal image, which they attempt to embody. According to Žižek, to be a subject, one must "put [one's] identity outside [one]self" (Žižek 116) into the image of one's double. Yet this relation between one's self and the alien image may be revised in encounters with its constitutive materiality. To this end, we highlight Burns' trope of messiness: repeated, artful messes such as thickets and bohemian interiors proliferate in panels and spreads, featuring expressive clues and detritus in contiguous flows. This trope runs through the comic and illustrates the inescapability of both ecological embeddedness and change, plus the inescapability of Burns' characters' participation within their environments. Burns' techniques bring characters into contact with their embedded status amid a wider ecology, suggesting social structures can be viewed as ecological—with both natural and societal networks liable to the vulnerabilities of contamination and entropy.

The Bug: Toward an Eerie Virality

Critical consensus informs our treatment of the Bug as a viral, rather than a bacterial infection, though Burns leaves its status deliberately ambiguous. Our analysis recognizes the inevitability of historical comparisons between the Bug in *Black Hole* and the HIV/AIDS crises of the 1980s, as documented by various critics and reviewers (Nandi and Parui; Jacobs; Zeigler; Raney). Such comparisons may be fruitful, as in Jacobs' argument that Burns "echoes the anxiety around HIV/AIDS in the early 80s" when the disease spread rapidly, but "little scientific explanation was available" (Jacobs 54). However, we wish to problematize any literal, or inflated comparisons between the Bug and HIV/AIDS, such as those offered by Raney, arguing that the Bug "stands in as a metonym for the AIDS virus" and that "AIDS became a metonym for deviance, for otherness" (Raney 2005). Our study seeks to problematize these comparisons due to their offensiveness in negating the individual lived experiences of HIV-positive people. Nevertheless, such readings inform our treatment of the Bug as viral. Similarly, Burns' artwork is redolent with visual motifs of viral pathogenesis in the form of compromised boundaries: incision-style wounds, torn skin, mutated orifices, broken bottles, gaps in forests, among others. Such visual breaches mirror viral reproduction, as viruses multiply by breaching cellular integrity (Drexler).

Typically, viruses are narrativized as invaders or intruders from the outside that threaten human autonomy by changing or controlling their biological systems. *Black Hole* resists this narrative and instead depicts the virus as part of the teen's lives, intrinsically linked with them post-infection. Significantly, the Bug is only perceived through its effect on the character's bodies; it is never pictured directly. Rather than being reductively narrativized as an intruder entering a pristine interiority, the Bug shows agency in a more embedded manner. The Bug destabilizes, distorts, and warps the characters' physiology, recalling their participation in a multispecies ecology, thus rendering the inside/outside distinction untenable.

The characters' mutations are an "irruption" (Fisher 20) of the outside Fisher argues is constitutive of the weird, radically altering the characters' sense of subjectivity and the understanding of symbolic orders associated with it, replacing discrete personhood[3] with a more consciously entangled state of being.

The Bug likewise belongs to another of Fisher's aesthetic-ontological categories: the "eerie" (Fisher 63), a phenomenon which in contrast to the weird concerns itself with "perhaps the most fundamental opposition of all-between presence and absence" (Fisher 61). While the Fisherian weird is synonymous with "an exorbitant presence… that exceeds our capacity to represent it" (Fisher 61), the eerie is marked by the "failure of presence," or "absence" (Fisher 61). This tension leads Fisher to argue that behind all manifestations of the eerie "is the problem of agency" (Fisher 63), giving examples of unpeopled landscapes, or an "eerie cry" (Fisher 11) emitted by an unknown creature, "what kind of agent is acting here? Is there an agent at all?" (Fisher 11). Despite Fisher's careful distinction between the two modes, viruses are an example of an entity that can be both weird and eerie. Importantly, viruses immanently unsettle dualisms of: presence and absence; living and non-living matter and that of self and other. Multiple studies by virologists have argued that "as there is no generally accepted definition of life or "aliveness" (Koonin and Starokadomskyy 131), such definitions are "neither feasible nor needed" (Koonin and Starokadomskyy 132). E.V. Koonin and P. Starokadomskyy note that to answer the question of "are viruses alive" "depends on the definition of life or the state of 'being alive,' and any such definition is bound to be arbitrary" (Koonin and Starokadomskyy 132). Therefore, viruses occupy a liminal "eerie" space between discrete biological and metaphysical categories. Furthermore, while no "deliberative agency" (Fisher 65) is possible with the Bug, its agency is expressed in the narrative through its symptomology, as is the case with extant viruses. The symptoms are visible, yet the microorganism itself is not.

The Bug, like existing viruses, significantly alters the behavior of characters, such as Keith, Chris, Rob, Eliza, and Dave, their relations to one another, and to the symbolic orders they inhabit. Narratively, however, the text downplays the significance of its mutagenic symptomology. Jacobs notes that "nobody knows" what "the long-term effects of [the Bug's] mutations might be" (Jacobs 54), and Burns omits scientific explanation for the Bug, with its pronounced mutations receiving no attention from medical authorities, even being a topic of Todd's adolescent derision. The Bug's presence is expressed in its effects, yet never depicted itself. The position of the Bug within Burns' graphic novel recapitulates the difficulty in placing viruses on either side of the living/non-living binary, with its ambiguously absent presence disrupting the binary of self and other. One way that the Bug exerts this eerie agency is the possibility of its compromising the human characters' agency. Relatedly, Emma Jacobs argues that in *Black Hole*, there is "an intoxicating, irresistible force of lust which guides' most of the story's major plot turns" (Jacobs 59). Although Chris is initially unaware of Rob's disease, she senses something pulling them together that is "dark and sexy" and "crazy and dangerous" (Burns). This unseen, "dark," "crazy," "dangerous" agent of lust is what the viral vector uses for its mode of transmission. While the Bug is not shown to affect the characters' behaviors outright, the presence of the Bug makes it harder to maintain the fantasy of absolute human autonomy. The Bug, occupying an eerie space between material and non-material, impinges also on the characters' decisions and behavior.

The Bug's absent presence is visually represented through the mutations of the characters. In this sense, the Bug is eerie, but that eeriness creates a weirdness; a sense of "wrongness" (Fisher 15). The mutations it causes and the character's animalistic features are in a place where they "shouldn't be" (Fisher 15). The "wrongness" (Fisher 15) of an eerie actant that cannot be seen demonstrates that not all actants are visible, undermining the assumption that only humans have agency. In compelling characters to acknowledge nonhuman agency, the Bug's "wrongness" indexes its capacity to disrupt treasured interiorities and catalyze alternative positions: this is what we term the productive abject.

Not Only Horrific: The Abject as Transformative

What this chapter terms the productive abject is the initially revolting encounter with that which subverts the self/other distinction, yet whose integration is potentially transformative, opening the space for a counter-hegemonic politics. The most influential text on the abject, Julia Kristeva's *Powers of Horror: An Essay on Abjection*, defines the abject as concerned with the "primal division" (Kristeva 10) between self and other. The abject is that which is "radically excluded" (Kristeva 2), yet necessary for the self-definition of an entity, whether an individual or community. The source of horror in abjections, for Kristeva, stems from the continued presence of this excluded matter beyond a given symbolic order, as the abject "preserves what existed" before the formation of that order; the "immemorial violence with which a body becomes separated from another body in order to be" (Kristeva 3). As Kristeva argues:

> There looms, within abjection, one of those violent, dark revolts of being, directed against a threat that seems to emanate from an exorbitant outside or inside, ejected beyond the scope of the possible. [...] It lies there, quite close, but it cannot be assimilated [...] a vortex of summons and repulsion. (1)

For Kristeva, the subject achieves its autonomy through rejecting what is considered improper or abnormal—this "exorbitant outside or inside' (1). Judith Butler maps the political and ideological dimensions of this process, and its implications for subaltern groups: the subject "gains its specificity by defining itself against what is outside itself" (Butler 142). Subjection to such norms involves rejecting elements of oneself as "spectres of abjection" (Butler 142).

Critically, as with Kristeva, the abject has been influentially conflated with the tropes, plots, and affects of the horror genre to an emphatic degree. The abject has also been evinced by horror scholars as a pronouncedly hegemonic, as opposed to counter-hegemonic, feature in many examples from the genre. For instance, feminist film scholar Barbara Creed identifies figures of danger in several horror films that are threatening to masculine spectators, arguing that the Hollywood horror film "stages and re-stages a constant repudiation of the 'maternal figure'" (Creed 70), which is operatively both abject and misogynistic. Conversely, the repudiation of grotesque matter such as viscera and mutated flesh in *Black Hole* is often transformative, precisely in terms of a disavowal of patriarchy and conservative social mores. In one segment of the comic, Chris hides in the girls' toilets at school while feeling stressed about the problems she is experiencing with her disingenuous love interest, Rob, with whom she has recently had unprotected sex. She muses about the girls chatting outside: "Those girls out there… how can they do it? How can they play along with all this shit?" (Burns). This moment mirrors with an early scene when Chris, surrounded by a reassuring cocoon of objects (a candle stub, socks, sneakers, paperbacks, Marlboros, uneaten food, Tampax), discovers there is a pliable tear in her skin and proceeds to pull her skin from her body before flinging the eerily disembodied epidermis away from her in contempt. Chris' peer and sympathetic admirer Keith finds her rejected skin in the woods: a compelling piece of evidence, not of Chris' repudiation of her own body or her sexual behavior, but of the outermost "page" of her presentation—her skin—which stands in for her "thrown" and presented self. In Žižek's terms, Chris' skin literalizes her "identity outside [one]self" (Žižek 116)—a claustrophobic, false subject borne from social necessity. Likewise, in an ironic circuit, textual reproductions and paper communicate in *Black Hole*, gesturing toward patriarchal symbolic orders. For example, Chris abandons her skin in the woods just as another character discards a Playboy magazine which is then found by Keith and his friends. In both cases, the papery items intimate toward the cultural embeddedness of the characters in systems of constricting and patriarchal symbolic reproduction and Chris defies this system of textual situatedness, inside of which she is objectified as other, by casting off the 'outer page' of her skin in the forest and beginning anew.

Significantly, not all visitations of the abject are horrible in nature or service the creation and reception of horror. In his work on queer male relationships, David Halperin states that, after the HIV/AIDS crisis, infected men "capitalised" on their "experience of abjection" "to think about their own bodies not just as bearers of contagion, not only as poisoned or ruined, but also as sparkling with viral light, as harbingers of a new ecology" (Halperin 86). The nature of expulsion here is not a moment of repulsion or devastation but instead transformative, allowing a re-imagining of viral infection. What Halperin describes is an example of the productive abject: a reclaiming of the abject to re-envision identity and one's position within wider symbolic orders. One might argue that the characters of Eliza and Keith embody these ideas in particular via their character arcs. As we will explore, the naturist artist Eliza uses her work, which investigates the taboo and abject, as inspiration for a life that avows the immanent, spontaneous, and unashamed as opposed to the preconditioned. Her monstrous and animalistic artistic creations also challenge humanity as a hermetic symbolic zone, upending rigidly hierarchized anthropocentrism.

The abject in *Black Hole* has been analyzed by Kom Kunyosying in terms of the graphic novel's formal and aesthetic ecology. Kunyosying foregrounds the role of Burns' "deliberately ambiguous significations" (Kunyosying 570), where Burns' art occupies a "liminal space between the human and the abject" (Kunyosying 570). This liminal space enables Burns' objects to escape containment under concepts that would reduce them to an "anthropocentric symbol" (Kunyosying 572). Kunyoysing examines the text's "formal engagement with ecology" (Kunyosying 570) and more specifically the ecological abject. The subject renders nonhuman nature and ecological embeddedness abject by violently making it the constitutive other of subjectivity. Rebeca Weiman notes that abjection creates a sense of "the more-than-human" (Weiman 7), which "governs my relationship with … non-human entities" (Weiman 7). Kunyosying's detailed study of the ecological abject present in *Black Hole* foregrounded our own explorations of productive, counter-hegemonic facets in the work. The "more than human" (Weiman 7) quality of abjected ecological embeddedness up-ends attempts to dichotomize humanity and nonhuman nature. Weiman argues that nature possesses the power to destroy the subject both physically and symbolically, "dissolving its boundaries or making it abject" (Weiman 20). This dissolution of rigid categories by abjected embeddedness is one of abjection's counter-hegemonic potentials. In *Black Hole*, Burns evokes multiple instances of visual and textual synonymy between anti-ecological symbolic orders and fast food, among other detritus. In one of Chris' flashbacks, she reflects on the weirdness of the Herfy's diner logo: "A big, dumb cow ready for the slaughterhouse … It's like he's saying" before completing her sentence out loud, "eat me" (Burns). Chris' identification with the cow and its invitation implies a synonymy between her position as woman under patriarchy, and the exploitation of nonhuman animals under capitalist agriculture. Despite gendering the cow as masculine, her affinity with the cow points to how both of them, placed within certain symbolic orders, are consumable and disposable. One of Chris' early nightmares sees her recoiling from her classmates, who are chest-deep in "floating garbage, French fries, old burgers" (Burns) that their bodies have become coextensive with. The discarded, processed food the teens uncritically eat is one such "deliberately ambiguous" signification (Kunyosying 570), positioning the teens in relation to the waste-matter that the symbolic order of capitalism produces. The stylized cow wanting to be eaten and the teens eating the garbage are powerful images of the cyclical nature of uncritical consumption of norms. Post-infection, however, Chris changes her attitude, empathizing with the mutated teens eating from the garbage despite her ability to "pass": "I was one of them. I just didn't show as much" (Burns). This "awful reality" (Burns) evinces Chris' anxiety about her future—the same situation that renders the infected teens disposable.

Another example of the transformative nature of abjection is expressed in Keith's character arc. In "Bag Action," Keith leaves his machismo-fueled friends, Dee and Todd, for the woods. While in the bathroom, Keith hears Jill protesting how little she enjoys the one-sided sex with Dee, asking out loud, "What am I doing here? Why did I think this was going to be fun?" (Burns), indicating Dee's unbearable patriarchal treatment of Jill in their relationship. Later, Keith's discomfort is realized in a large panel spread. His body is distended in front of a shop window, stocked with commodities, which he refers to as "the most ungodly shit imaginable" (Burns). Burns warps and stretches the visual signifiers of American suburbia repeatedly in *Black Hole*—signifiers such as white picket fences, family portraits, kitsch commodities, and yearbook photos. However, this scene demonstrates an augmentation of Keith's consciousness that allows him to experience the constricting character of patriarchal gender and class norms via their abjection. By comparison, when his journey leads him to confide in a group of infected teens in the woods, Keith is able to verbalize his discomfort with such norms: "As I started to talk, I felt something swelling in my chest, straining to get out … It came spilling" (Burns). Likewise, Keith finds solace in the group of abjected teens: "some normal, some distorted, but all glowing with a warm internal light" (Burns). Keith's journey to the group demonstrates the capacity of the productive abject to render boundaries porous, enabling passage to new forms of expression and new modes of existence achieved through the rejection of bodily and gender norms.

The Productive Abject, Boundary Fluidity, and the Grotesque

Black Hole avoids falling into the hegemonic horror described above by Creed as its abject encounters are productive. By this, we mean that abject encounters—either horrible, revolting, or with the potentiality to cause enigmatic rupture—subvert dualisms by bringing something new, unwelcome, and weird unexpectedly into a frame of reference. Their introduction generates new possibilities for individual coexistence and, relatedly, renewed social solidarity. Part of the weirdness of Burns' graphic novel lies in his subversion of distinctions and the cycling demonstration of the porosity of boundaries—a point made in the valences of his plot as much as with the flux and vacillations of his simultaneous objects. Weird events disrupt symbolic orders in his plot but the effect of this weirdness, the productive abject, is instantiated in Burns' visual registers. While the assumed reaction would be that characters recoil in a superior sense of revulsion at the mutations Burns depicts, they also evoke empathy for the teens' former enclosure in the strictures of suburban American life. In terms of the "products" of abjection, evidence abounds in the work of the characters' almost universal, deep unhappiness and alienation at the outset of the comic series and this alienation is disrupted by the narrative's unusual events. Individual and social stagnancy is penetrated in the work, with some momentum, by the abject and weird. This includes an opening scene in which, in a haruspicy-type psychological event, the sensitive character Keith sees an infernal, messy, yet undeniably exhilarating future whirling in the guts of a dissected frog. This is a future that Keith both fears and welcomes.

In the sense of an inside-outside boundary and the penetration of revelatory phenomena, porousness is a constituting factor of Burns' use of the abject and weird. The traversing of this boundary in a biologically material sense and the consequent repulsion it produces is aligned with body horror. The horror that comes with the graphic depiction of bodily differences makes those changes grotesque. *Black Hole* operates in the genre of the grotesque—a kin genre of the weird. Fisher argues that the grotesque can encompass "human and animal shapes intermingled with foliage" (Fisher 32). The Bug literalizes this intermingling in the mutations it produces, such as Chris' snakelike skin, Eliza's tail, and Dave's leonine mane.

Returning to Chris' skin, this is the most striking example of Burns' use of the grotesque to destabilize existing symbolic orders. In the chapter "Seeing Double," Chris draws attention to the otherworldliness of her mutation, describing it as "soft and open," a "loose tear," and "not like a wound, not like anything" (Burns). Familiar images of skin tearing, such as wounds, are immediately rebuffed before Chris admits her mutation has no point of comparison, as it is "not like anything" and seemingly offers no pain, as it offers "no resistance" (Burns) to her attempts to explore it.

Chris' skin exemplifies the features of the grotesque as an evocation of "something out of place" (Fisher 32), while its being "not like anything" (Burns) demonstrates an affinity with ambiguous materialities that elide neat characterization. The skin acts as a literal boundary between self and other. Chris' skin shedding and tearing demonstrates the reality that the skin is a porous boundary. The breaking of the skin and its deterioration is a literalization of dualistic symbolic orders breaking down through the "irruption" of the weird. Chris' skin is a paradigmatic simultaneous object, with its texture and tears synecdochically connecting her with other objects: nervous systems, parchment, tree branches, or snake skin. Chris' mutation is an example of the grotesque in the precise definition Fisher offers: "human and animal shapes intermingled with foliage" (Fisher 32).

Burns further demonstrates the porosity of boundaries through his use of humanimality, troubling the rigidity of the human/animal binary. The boundary of the "inside" of human personhood is eroded and breached by the apparent externality of the environment. Cary Wolfe argues that the concept of the "human" is born from establishing a "humanity/animality dichotomy" (Wolfe xv), where the former is achieved by repressing or escaping the latter. For Wolfe, posthumanism comes both "before and after humanism" (Wolfe xv), evincing humanity's embeddedness not just in biological networks, but beyond in tools, culture, and networks with nonhumans. Therefore, the boundaries of the human are permeable at all levels. Packwood Freeman likewise provides a definition of "humanimality" for communications theory, which outlines how "humans might rhetorically construct themselves as animals" (Freeman 11) to engage with the complexities of animal rights, among other ecocritical phenomena. By making a perceived "outside" (Fisher 20) (the natural world) simultaneous with skin and nervous systems, Burns shows the biological embeddedness of humans within the environment. In *Black Hole*, the animal mutations that signify the Bug's symptomology also literalize the similarities between human and other-than-human DNA. In addition to this genetic consideration, as with the provenance of zoonotic diseases, humanimality can encompass consideration of geographical habitats as not solely for human consumption and occupation but as shared, and thus capable of contamination for all parties.

Rosi Braidotti argues that, as European humanism is problematically equated with "consciousness, universal rationality and self-regulating moral behaviour," these attributes of personhood are defined against "Otherness"—the subject's "negative and specular counterpart" (Braidotti 15). Such "imperial humanism" identifies difference with abnormality and, broadly, inferiority. Deviation from such norms renders populations subject to them as abject in various ways, as when working class, gendered subjects, racialized subjects, or disabled subjects are framed as counter-normative and antagonistic to dominant symbolic orders. Notably, Burns' Bug and its symptomology illustrate the connection between the human subject and these "disposable others" (Yazgünoğlu 46). In *Posthuman Pathogenesis*, Kerim Can Yazgünoğlu discusses the posthuman nature of viral infection: "Nonhuman actants in our bodies will constantly remind us of the eroded boundaries between our lives' and the 'disposable others'" (Yazgünoğlu 46). In the case of *Black Hole*, the "disposable others" (Yazgünoğlu 46) are the natural world, women, and infected people. This message is particularly emphasized via the aggressive figure of Dave, who succumbs to a leonine mutation as alienating as it is totalizing, as it causes his entire face to appear lion-like, making him unrecognizable. Dave, far from a popular favorite in his school days, cannot cope with his increasingly marginal status after infection and eventually succumbs to violence and suicide: this scene is attended by the fetching and consumption of a chicken bucket, which Dave steals at gun point. Kunyosying writes,

> Hageman sees significance in the environmentally irresponsible greasy food at Winkie's and analyzes it as ecologically relevant because it underscores the unswallowable nature of the scene. … The bucket of chicken used to set up a murder in *Black Hole* contains ecological commentary on fast food … Dave brings his closest friend Rick the chicken to give him one last treat as well as distract him from what Dave sees as his mercy killing. … Throughout *Black Hole*, discarded fried chicken bones and buckets are strewn among detritus and scattered flora and fauna. As products of both humans and animals, the chicken bones occupy a space between the abjection of human waste and the abjection of the ecological as nonhuman. (Kunyosying 569)

The chicken detritus that attends Dave's narrative arc speaks of his violent behavior as symbolically associated with human beings' environmental devastation, insofar as those human beings collude with damaging systems. The chicken imagery doubles with a notion of Dave and his friend as societal castoffs, unworthy of full protection, and this is reminiscent of Yazgünoğlu's posthumanist theory connecting the abjection of nonhuman agency to how subaltern populations become "disposable others" (Yazgünoğlu 46) within given symbolic orders.

As we have seen, Burns' Bug literalizes an "outside" (Fisher 20) of an often mute environment, along with threatening forms of social precarity, emerging on the "inside" (Fisher 20) of bodily materiality and the characters' formerly ordinary lives. Through bodily mutation plus coterminous experiences of social othering, Burns illustrates how his characters are abjected. Their mutation is a process of becoming other, as Burns' characters fantastically merge with the natural world. These animalistic mutations literalize Freeman's notion of "humanimality" (Freeman 11), undercutting the human/animal distinction Wolfe and Braidotti both argue is foundational to the category of "human." This throws away, like Chris' old skin, the symbolic framing of natural actants as something "over there."

Breaking Symbolic Orders and Beginning New Imaginaries

In *Black Hole*, two of the most important characters that elucidate the productive abject in action are Keith and Eliza. There is scope for further research on how all the characters in the graphic novel are exposed and react to the productive abject, yet for the scope of this study, we argue that Keith and Eliza are the most prominent and worthy of our explication here. The externality of the productive abject enables an ideological critique of the symbolic orders through which subjects are rendered "disposable others" (Yazgünoğlu 46). Keith's narrative arc is typified by a stasis legible in terms of the symbolic order of patriarchy. His machismo-fueled friends, Dee and Todd, affirm the presence of stultifying gender dynamics in their chauvinistic mannerisms and worldview: smoking dope in exclusive groups of men and treating women as contemptible and functionally identical. Pre-infection, Keith suffers from an uneasy, oscillating relationship with these patriarchal gender norms, and Todd calls Keith a "martyr" (Burns) for being concerned for his crush Chris. After meeting Eliza, Keith views these behaviors as social toxins he wishes to extricate himself from—for example, when he is unable to comprehend why Eliza would "live with a bunch of burned out college dudes" (Burns). Their solipsistic practices are no longer desirable: in fact, Keith increasingly views them as socially damaging, comparable to the ecologically damaging effects of the trash surrounding his former friends, Dee and Todd, in "The End" (Burns).

In "The End," Burns' art emphatically realizes the implied synonymy of patriarchy with an insulated adolescence, aloof to its interconnection with others, and ecological pollution. Keith's nightmare recalls friends Dee and Todd discovering Rick's yearbook in his tent—a peer they ignored and perhaps picked on before infection. The nightmare yearbook depicts the teens in various stages of mutation, and Dee derisively comments, "Look! It's got pictures of all your friends!" (Burns). When Keith sees an image of the first infected teen he encountered in the woods, Todd exaggeratedly mocks him: "It's you, ya big dork!" (Burns). This moment consolidates the defamiliarizing nature of the productive abject as Keith no longer recognizes himself, and this nightmare affirms that the anxiety and dissociation Keith repeatedly suffers from is one signal of many that he finds his close friends and their derision stultifying and alienating.

Keith's infection and consequent passage to social disposability cohere with an embrace of the weird irruption of the Bug into his life. This irruption also causes him to identify his former habits with pollution, finding them among detritus at the bottom of a ravine full of "endless garbage… glass, shit, chunks of styrofoam, plastic forks, torn up magazines, candy wrappers, cigarette butts, crumpled beer cans" (Burns) in his nightmare. Keith's nightmare exemplifies the potential of the productive abject, in the sense that the horror of departing from a hegemonic position gives way to an understanding of the innate harm of reactionary or retrograde symbolic orders. In Keith's arc, the text discloses the mutual violence of patriarchy and anthropocentrism by first denaturalizing them, aesthetically rendering their co-extension with forms of subjectivity through the merging of bodies and trash. This example of Burns' contiguous flow technique re-classifies negative symbolic orders as deleterious.

In a related vein, Burns' treatment of the artist Eliza depicts the expulsion of non-conforming women from existing symbolic orders plus the expulsion of new ontologies that they might represent. Unlike other carriers of the Bug, Eliza at first appears to have embraced her mutation, given the playful connotations ascribed to her tail "moving just a little bit…twitching" (Burns). Eliza's artwork is also posthumanist, featuring a blending of human and animal figures, such as frogs with human eyes, caterpillar necks culminating in human faces, and human figures with insectoid or crab legs growing from their heads. In her comparative easiness with her mutated state, Eliza and her artwork illustrate the productive valences of humanimality and posthumanism. Eliza's work demonstrates the porosity of boundaries between two apparently distinct realms. Burns' illustrations of Eliza's art gradually take up more panels, eventually having whole segments dedicated to them: a figure with legs emerging surreally from its head and a bound male figure in the woods in particular dominate. Thus, not only do Eliza's artworks upend the "hierarchical scale of decreasing worth" (Braidotti 105) associated with humanistic ideologies, but also they effectuate a space of porosity both between humans and other animals and between content and form in *Black Hole.* Eliza's playful embrace of the Bug's mutations and the wider field of material interconnection thus make her into a self-aware figure of the cyborg. As theorized by Haraway, the cyborg is "a condensed image of both imagination and material reality" that disrupts the false dichotomies "between mind and body, animal and human, organism and machine, public and private, men and women, primitive and civilized" (Haraway 32). Embracing the Bug, Eliza is content living between the poles of animal and human rather than forcing herself to comply with them. Her animal-human artwork demonstrates how her imagination (as well as her material self) becomes hybridized through processes of liberating self-awareness.

Eliza embraces the abject in her art, surrounding her bed with figurations of both freakishness and the sometimes violent punishment of aberrant behavior and identity. Attesting to their productive status, Keith fully embraces her artworks, remarking, "it's great!" (Burns). Keith, viewing Eliza's art, is exposed to a novel form of self-knowledge: one that embraces difference and unflinchingly portrays the injustice and emotional consequences that attend systemic violence. Keith internalizes the messages of Eliza's work and subsequently outgrows his previous understanding of the world formerly entrenched by Dee and Todd. This moment of exposure to the productive abject propels his journey to relinquishing his overbearingly patriarchal perspective, principally manifested in his unrequited obsession with Chris. Instead, Keith embraces a more thoughtful relationship with Eliza. The tadpole nature of Keith's later mutations gesture in a tongue-in-cheek manner to his developing awareness of the initial frog in "Biology 101," which divined his weird future. Keith's character development regarding a disavowal of patriarchy is partly tracked by ironical imagery regarding frogs in the narrative, as the wound in the dissected frog is graphically reminiscent of a vulva and gestures to Keith's immature masculine anxiety. In later scenes, the comparatively inoffensive and painless tadpoles that grow from his skin gesture backward to the dissected frog, yet suggest the beginning of a new cycle toward open-mindedness as opposed to fatalism in Keith's attitudes toward women. Through the imagery related to his premonition and then his mutation, Keith begins a productive journey toward a more self-aware and socially conscious understanding of the world, compared to his previous discriminatory approach to feminine lust and love objects.

In terms of patriarchy's hold on other realms of Burns' narrative, the violent suppression of Eliza in the story is also linked to her sexual confidence, self-expression, and non-compliance with pre-set notions of womanhood. Eliza's housemates react violently to her acceptance and embracing of her infection along with the other counter-hegemonic facets of her character. Burns abjects Eliza's assault in the sense that it is painfully foreshadowed and then described from Eliza's perspective. At first, Eliza's housemates refuse to clean their shared environment and her bedroom door is defaced with crude, sexualizing graffiti. Finally, men at their house party (along with, presumably, Eliza's male housemates) drug Eliza, write and draw on her body and then rape her, violently overdetermining her non-compliance with symbolic orders by asserting their own worldview literally onto her flesh in multiple ways. The inscription of Eliza's inert body doubles for the violent resistance of dominant symbolic orders toward new ontologies. Her housemates' crude graffiti is the darkly satirical and undermining double of Eliza's unusual outsider art. The physical and semiotic violence within this act of corrective rape demonstrates an attempt to impress Eliza back into subordination, potentially as a violent response both to her infection and to her failure to abide by other 1970's gender norms. Eliza's counter-hegemonic behaviors include her love of nudity and her former trust of, and desire to live alone with, a group of men.

This chapter detailed the corrosion of anthropocentric symbolic orders in Charles Burns' *Black Hole*, as well as his aesthetic and ethico-political strategies for heralding an ontology of coexistence. In his visually recurrent messes, and pictorial co-extension of the human body with those of other animals and their environs, Burns' text foregrounds a narrative ambiguity that renders stable symbolic categories permeable. Without explicit authorial intention, *Black Hole* provides a much-needed corrective to ideologies, which deny the validity of multispecies coexistence and anthropogenic climate change. Its aesthetic and character subversions are appropriate to a post-dualistic, critically posthumanist worldview, with the force acting upon the registers of aesthetic and character psychology being what we term productive abjection. The productive abject is a similarly posthumanist aesthetic and ethico-political process, as it both reveals invisible boundaries and provides the means of corroding them. *Black Hole* discloses ecological embeddedness and necessitates revising ontological commitments fraught with blindness to social othering and alienation. The text itself lacks concrete ethico-political alternatives: the implication of the Bug's future consequences haunt Keith and Eliza's blossoming relationship, as well as accelerating Chris' growing existential isolation. Nevertheless, in its weirding of current symbolic orders, *Black Hole* occupies a point of rupture, tearing a hole in the ideologies of anthropocentrism.

Notes

1 See Plowright et al. for discussions about zoonoses spreading from horses to dogs and humans as an example.
2 This chapter analyzes the Jonathan Cape 2005 edition of *Black Hole*, which lacks page numbers. Hereafter, all references to *Black Hole* are indicated by a parenthetical reference to the author's name.
3 For a detailed summary of the link between ideologies of personhood and social domination, see Braidotti, *The Posthuman* (2012).

Works Cited

Braidotti, Rosi. *Posthuman Knowledge*. Polity Press, 2019.

———. *The Posthuman*. Polity Press, 2013.

Burns, Charles. *Black Hole*. Johnathan Cape, 2005.

Butler, Judith. *Bodies That Matter: On the Discursive Limits of "Sex."* Routledge, 2015.

———. *Frames of War*. Verso, 2016.

Drexler, M. *What You Need to Know about Infectious Disease Institute of Medicine*. (US) National Academies Press, 2010. https://www.ncbi.nlm.nih.gov/books/NBK209710/

Fisher, Mark. *The Weird and the Eerie*. Repeater Books, 2016.

Freeman, Carrie Packwood. "Embracing Humanimality: Deconstructing the Human/Animal Dichotomy" in Goodale, G and Black, J.E, editors. *Arguments about Animal Ethics*. Lexington Books, 2010.

Halperin, David. *What Do Gay Men Want? An Essay on Sex, Risk and Subjectivity*. U of Michigan P, 2007.

Haraway, Donna Jeanne. *Manifestly Haraway*. U of Minnesota P, 2016.

Jacobs, Emma. "The Air Is Pure Poison: Precarity and the Pseudo-Apocalypse in *Black Hole* and *David Boring*." *English Academy Review*, vol. 31, no. 2, 2014, pp. 49–63. doi: 10.1080/10131752.2014.965416.

Koonin, Eugene V., and Starokadomskyy. "Are Viruses Alive? The Replicator Paradigm Sheds Decisive Light on an Old but Misguided Question." *Studies in History and Philosophy of Biological and Biomedical Sciences*, vol. 59, 2016, pp. 125–34. doi: 10.1016/j.shpsc.2016.02.016.

Kristeva, Julia. *Powers of Horror: An Essay on Abjection*. Columbia UP, 1980.

Kunyosying, Kom. "Horror Comics Ecology: Metonymy and Iconicity in Charles Burns' *Black Hole*." *ISLE: Interdisciplinary Studies in Literature and Environment*, vol. 21, no. 3, 2014, pp. 562–74. doi: 10.1093/isle/isu084.

Nandi, Arindam, and Avishek Parui. "'And Then It Spreads': Contagion and Disease as Metaphors of Sociomoral Contamination in Charles Burns' Graphic Novel *Black Hole*." *Medical Humanities*, 2023, doi: 10.1136/medhum-2023-012625.

O'Callaghan-Gordo, Cristina, and Josep M Antó. "COVID-19: The Disease of the Anthropocene." *Environmental Research*, vol. 187, 2020, p. 109683. doi: 10.1016/j.envres.2020.109683.

Pizzino, Christopher. *Arresting Development: Comics at the Boundaries of Literature*. U of Texas P, 2016.

Plowright, Raina K., et al. "Ecological Dynamics of Emerging Bat Virus Spillover." *The Royal Society*, 2015. https://doi.org/10.1098/rspb.2014.2124

Protic, Nemanja, and Niel Finlayson. "Black Holes and Parallax Gaps: The Contemporary Graphic Noir of Charles Burns' *Black Hole*." *Journal of Graphic Novels and Comics*, vol. 7, no. 1, pp. 3–19. doi: 10.1080/21504857.2015.1093517.

Raney, Vanessa. "Review of Charles Burns Black Hole." *ImageTexT*, 2005. imagetextjournal.com/review-of-charles-burns-black-hole/.

Weiman, Rebeca. "Dark Ecology and the Abject." *International Journal of Žižek Studies*, vol. 10, no. 3, 2014, pp. 1–22. ISSN 1751-8229.

Wolfe, Cary. *What Is Posthumanism?* U of Minnesota P, 2010.

Yazgünoğlu, Kerim Can. "Viruses as Posthuman Biocultural Creatures: Parasites, Biopolitics, and Contemporary Literary Reflections." *Posthuman Pathogenesis: Contagion in Literature, Arts, and Media*, edited by Basak Agin and Safak Horzum, Routledge, 2023, pp. 42–59.

Žižek, Slavoj. *The Sublime Object of Ideology.* Verso, 2009.

11 (Un)Caring Borders

More-Than-Human Solidarities in the Białowieża Forest

Hanna Grześkiewicz and Marleen Boschen

Introduction

A new 5.5-meter tall border wall enforced with razor wire and an electronic barrier runs through the ancient Białowieża Forest, marking the outer edges of the EU along Poland's Eastern border. It is a barrier built to stop those passing through the forest from Belarus to Poland, with the goal of finding safety in the Western EU states. The people on the move making this journey come from Afghanistan, Iraq, Syria, the Democratic Republic of the Congo, Yemen, and many places where war and other crises have forced them to seek safer destinations. The wall was erected in the midst of emergency regulations that for some time turned the Polish side of the Forest into a zone inaccessible to the public. The Białowieża Forest is Europe's last primeval forest, made up of ancient trees and home to a plethora of species such as mosses, lichens, fungi, and insects that depend on the deadwood of old trees as well as roaming mammals. At the time of writing, in spring 2023, the wall has been erected and its electronic fixtures are being finalized—it is fully operational along the majority of its length, equipped with motion sensors and a sophisticated monitoring system. The wall stretches for 186 km, and the electronic barrier reaches 206 km, operating also on sections where the border is demarcated by a river. Intended to stop people on the move who have been crossing the border since summer 2021, the wall—both physical and electronic—cuts through the forest's ecological communities.

An outbreak, not of the global COVID-19 pandemic, but of a bark beetle "invasion" by *ips typographicus* had catapulted the Białowieża Forest into headlines in 2016. At the time, the Polish State Forests organization initiated large-scale logging of affected trees, leading to international uproar and activist occupations of the Forest in attempts to protect its ancient biodiversity. Across these two conflicts, the current "humanitarian crisis"[1] and the earlier logging crisis, Białowieża's border ecology has been implicated in different national formations of care, vulnerability, and the othering of specific more-than-human communities. What makes these conflicts important to consider jointly is that discourses around outbreaks and "invasion" were instrumentalized. They were used to set up borders and boundaries between those human and other-than-human actors that are welcome and cared for, in the state's more-than-human body, and those who are not.

DOI: 10.4324/9781003345770-15

In January 2022, there were reports of a trapped herd of bison between the barriers on each side of the border. One bison was reported to have died, just next to the fence, the cause of death was speculated to be barbed wire injury (Łopienski). In April 2023, news circulated of a person on the move who fell from the border wall and subsequently died (Tomczak). At the same time, journalists reported on the inadequate openings for animal crossings built into the wall, with fears for bison, deer, lynxes, and other larger animals either getting injured, caught, or not being able to continue their migration path through the wall (Jurszo), which was meant to *only* deter humans. These reports of life trapped in the border fence create a violent image and crucially lived reality, which we work with to seek to understand the denial of care to humans and other-than-humans in Białowieża's border zone and the formations of solidarity and protest that emerged in spite of this. In this chapter, we ask: how do state authorities deny the agency of and care for more-than-human communities in the Białowieża Forest? And what formations of solidarity emerge in resistance to this?

We approach the Białowieża Forest from a distance, Marleen is a German researcher based in London and Hanna is a Polish researcher and activist based in Berlin. In 2019, Marleen conducted fieldwork with the Kostrzyca Forest Gene Bank, which belongs to the State Forests organization that oversees Polish forests and holds a DNA archive of protected species in the Białowieża Forest. Hanna has been part of the feminist, queer, and no borders movements around the Polish diaspora and locally in Germany. Throughout the pandemic, a visit to the forest was impossible for us and we acknowledge this tension of writing from afar about care.

Our access to the border crises is through international and Polish news reports and statements by state actors, international organizations, and activist groups. Access to information and language have focused our analysis on Polish(-language) sources. We are aware that this chapter's scope is shaped by reduced access to information from the Belarussian side.

A Brief History of Białowieża Forest

Since 2016, the Białowieża Forest, which stretches across the border in Eastern Poland and the Southwest of Belarus and is partially a UNESCO World Heritage site, has been a contested setting for divergent imaginaries of managing vulnerability, national nature, and invasion. These imaginaries are grounded in the othering of some more-than-humans, in a logging controversy from 2016 to 2018 and people on the move crossing the forest in an ongoing border crisis since 2021. In the spring of 2016, in response to a spruce bark beetle outbreak, the Polish Environment Ministry authorized the State Forests Department to start salvage logging of trees damaged by bark beetles in the unprotected area of the forest, discussed at greater length in the next section. This led to intense protests and occupations in parts of the forests as campaigners feared for the integrity of Białowieża's ancient and biodiverse ecology and fought for international protection of the site. At present, the UNESCO world heritage site on both sides of the border has a strict human non-intervention policy. On the Polish side, outside the UNESCO site, the forest stretches further across three forest districts—Białowieża, Browsk, and Hajnówka—which actively manage the Forest. It is in these managed parts of the Forest that the controversial salvage logging took place. On 17 April 2018, the Court of Justice of the European Union ordered the Polish government, led at the time by the right-wing Law and Justice party [*Prawo i Sprawiedliwość*, shortened to PiS], to immediately stop its salvage logging operation in the judgment of the case C-441/17 (2018).[2] The UNESCO status creates a zone of exception to national sovereignty as the EU Court of Justice decision opens an avenue for more-than-human sovereignty. Yet, the UNESCO status is not an embodied more-than-human expression of sovereignty, but an extranational, masterful sovereignty where "nature" is conserved and protected from human interference.

Enclosure of the forest-as-resource can be traced throughout the history of the Białowieża Forest, revealing afterlives of human intervention. The Forest has been used by humans since at least the fifteenth century on a commons-use basis (Samojlik 155–57). In 1812, Napoleon's troops passed through the Forest and used it as a hunting ground. Throughout the nineteenth century, Białowieża saw waves of large-scale logging until in 1888 it was officially protected as a hunting ground for tsar Nicholas II. During World War I, Germany extracted 4.5 million cubic meters of timber via a newly constructed railway, followed by a period of Polish governance of the forest that gave logging rights to the British company ETC, which extracted more than 1 million cubic meters annually (State Forests, "Frequently Asked Questions"). In 1939, the Forest became a Soviet nature reserve only to be occupied by German forces again during World War II. After the war, it returned to the status of a nature reserve on the Polish and Soviet sides. The histories hinted at here point to intervention in the Forest across periods of occupation and the different nationalist imaginaries at play during these stages of occupation. To what extent the ecological consequences of these occupations are still present in the forest is hard to say. As Macarena Gómez-Barris asks regarding Latin American forests, "what do we really know about the invisible, the inanimate, and the nonhuman forms that creatively reside as afterlives of the colonial encounter?" (xx).

The State Forests organization manages publicly owned forests in Poland and is financially independent from the Polish government. It is the biggest forest organization in Europe. It was established in 1928 and its holding totaled almost 30% of Polish territory (Blavascunas 35). In response to criticisms of logging in Białowieża, State Forests emphasized the restoration of past damage:

> The Białowieża Forest is a priceless treasure for foresters. For over 90 years, they have been striving to protect it and to repair all damages it suffered in the early 20th century. (…) If the foresters had not been persistent, many valuable parts of the forest wouldn't exist today. The Second World War was also tragic for the Białowieża Forest. In just two years of occupation, the Soviet authorities logged over 1.5 million cubic meters of timber. *Only after the end of war the nature, with the support of Polish foresters, could begin to heal the wounds made by the oppressors.* (State Forests n.p., italics ours)

It is worth analyzing the above statement more closely. First, the invasive threats to the forest were consistently foreign and extractive. Second, foresters were able to support the Forest's recovery. Third, paradoxically this history of extraction and injury to the forest is now used as a justification for further *caring extraction* in the logging conflict, this time of deadwood resulting from the spruce bark beetle. Here, the forester, as an extension of the state, becomes an aid to the Forest in a move that constructs the Forest as vulnerable and abused. Białowieża is transferred into a realm of national heritage, where "nature" is part of the national body.

Disturbances of the Forest

The ecological relations of the spruce bark beetle are destructive yet beneficial for ecologies, depending on the temporal scales of observation. It is thus important to trace how disturbance was instrumentalized by the Polish State Forests organization to legitimize the salvage logging of the forest and its cryo-preservation project. A careful analysis is needed of this "rhetoric of biological invasions" (Subramaniam 26). Our analysis looks at how otherness and calls for intervention are constructed in the Białowieża case in political readings of the situation that mobilize different temporalities. As Blavascunas suggests: "the forest mediates nostalgia and othering" (Blavascunas 8).

The European spruce bark beetle (*Ips typographus*) outbreak rapidly affected Norwegian spruce trees (*Picea abeis*) and, according to the Polish government, caused a dangerous disturbance to the forest ecology. However, ecologists have argued it was instead the logging operation that damaged the forest (Mikusiński et al. 267). The conflict that ensued between State Forests, ecologists, activists who occupied the forest, and the supranational bodies of UNESCO and the EU Court of Justice was an epistemic conflict around *productive* and *threatened* "nature," between the disciplines of modern forestry and conservation ecology. At the center of this were epistemically divergent discourses of disturbance, invasion, and intervention, which are equally echoed in the discourses around the humanitarian crisis.

We work with Anna Tsing's notion of "disturbance" to make tangible the more-than-human precarity of the Białowieża Forest on temporal and spatial scales. In her multi-sited multispecies study of matsutake mushroom pickers, Tsing observes "more-than-human sociality" in the relationship between pines, matsutake mushrooms, and their pickers. In a forest in Northern Finland, she describes what appears to be "an industrial tree plantation"—clean and homogenous, "both natural and artificial." Tsing explores what is lost in these approaches to the management of forests: "With modern forestry, we forget that trees are historical actors. How might we remove the blinders of modern resource management to regain a feel for the dynamism so central to the life of the forest?" (Tsing 168). Arguably, this ecological history and dynamism are what needs to be tuned into in the Białowieża conflict, specifically in how they relate to the framing of "invasion" (State Forests, "Spruce Bark Beetle"), and disturbance as the reason for the DNA collection project as a techno-ecological record of the Forest. Tsing uses disturbance to reveal different layers and ecological imaginations of invasion, restoration, and scientific objects of protection. Yet, State Forests' response to the spruce bark beetle "invasion" describes a different politics of disturbance, an ecological othering of invasive species. This instrumentalization of disturbance on a national scale, of ecological dynamism, is maybe something Tsing does not anticipate when she suggests that "restoration requires disturbance—but disturbance to enhance diversity and the healthy functioning of ecosystems" (Tsing 152) as it places the control of what counts as healthy functioning with those who hold ecological sovereignty.

A new layer of "disturbance" entered the forest in August 2021 when the authoritarian Belarusian president Alexander Lukashenko introduced visa-waivers as a means of offering people on the move a passage into the EU via the Belarusian-Polish border, kickstarting an entire economy and ecosystem of "travel" agencies, smugglers, activists, nongovernmental organizations (NGOs), and state actors. To conclude this discussion on disturbance "deciding what counts as disturbance is always a matter of point of view" as Tsing argues (161). The divergent discourses of disturbance—the spruce bark beetle as disturbance versus salvage logging as disturbance—present in the affectively charged Białowieża Forest reveal multiple paradigms of care and protection at play.

A Brief Timeline of the Humanitarian Crisis

In August 2021, the Polish town of Usnarz Górny made headlines when it was reported that 32 people from Afghanistan were being held in a clandestine camp, in inhumane conditions, surrounded by the military and barbed wire. This situation lasted approximately two months until reports stated that some escaped in mid-October 2021—some were captured again by the Polish Border Guards, but the fate of only some is known. Usnarz Górny is often marked as the beginning of the "humanitarian crisis" on the Polish-Belarusian border. Between 2 September 2021 and 30 June 2022, various restrictions were introduced in the border area on the Polish side affecting rights of access, movement, and assembly. For some time, the area near the border was declared to be "a state of emergency zone" [*strefa stanu wyjątkowego*], resulting in an entry ban for non-residents, which was especially directed at activists, lawyers, medical workers, and journalists. To this day, the presence of Border Guards and the military in the border regions is heightened. As analyzed by *Badacze i Badaczki na Granicy* [Researchers on the Border], the daily lives of local residents have changed significantly over the last two years.

While not wanting to reduce the humanitarian crisis to a set of numbers, they can help understand the situation. In their publication from December 2022 to January 2023, the activist network *Grupa Granica* [Border Group] reported having helped 14,500 people between August 2021 and January 2023 and listed the number of known missing persons to be 225, confirmed deaths to be 37.[3] The number of persons attempting to cross the border, as well as the numbers of those who are missing and those who have died, are likely to be significantly higher. There is no evidence that the newly erected border wall has been "successful" in reducing the flow of people—on the contrary even right-leaning publications report on its failure to do so (Chomiuk). It is important to highlight the difficult conditions in the Forest for humans, especially in the winter months. The border area is a deeply forested wetland, populated by rivers and lakes, meaning that the natural conditions are extremely precarious. This is in addition to the new border wall and electronic barrier, heavily armed military, Border Guards patrolling the border, and legal frameworks introduced by the Polish government that effectively allow pushbacks, often accompanied by physical and verbal violence (*Rozporządzenie MSWiA*).[4] Justifications of state actions on the border continue to be published through the government's publicity machine: on the official TikTok account of PiS, a spokesperson said "we are not condemning people to death at the border, we are defending Polish society" (@pis_org).

Crucial moments in this timeline happened in dialogue with COVID-19 pandemic restrictions. The pandemic allowed the Polish government to restrict the movement of people, for the first time since Martial Law was introduced in 1981, paving the way for additional restrictions in the border region. This unprecedented situation was a key reason for the ease with which policies around the border could be introduced.

Enclosures

Piro Rexhepi uses the term *white enclosures* (2023) to describe the European Union, a region heavily investing in militarization and fortification in order to seal its boundaries to protect the "fantasy of whiteness" (Mbembe 45). In activist spaces, this is commonly known as "Fortress Europe." Rexhepi argues that Eastern European countries have become a peripheral extension of European coloniality by formulating and enacting racist policies with regard to the border crossing of people on the move (Rexhepi 12). This extension into the peripheries of Eastern Europe is clear from the support of Frontex, the European Border and Coast Guard Agency, for the measures being taken at the Polish-Belarussian border. In a press release from October 2021, it was written that its chief Fabrico Leggeri

> was impressed by the means deployed to secure the border. He also thanked Poland for cooperating with Frontex since the beginning of the crisis through an ongoing exchange of information and providing the Agency with data on the situation on the Polish section of the external border. ("Frontex Executive Director…")

Even though the Frontex headquarters are situated in the Polish capital, Warsaw, unlike in Lithuania where the Eastern EU border with Belarus continues, the Polish government did not call on the support of Frontex. Paradoxically, this was used by the Polish government to propagate its Euroskepticism, celebrating that Poland can "defend its own borders," all the while the Polish Border Guard was "defending" the EU's Eastern boundary—and despite, as evidenced by Kamila Fiałkowska, Polish Border Guards having been trained by Frontex to ensure "uniform border control standards" ("Główne zadania", own translation).

The racialized nature of the policies at the Polish-Belarusian border becomes clear when compared with the arrival of those fleeing the full-scale Russian invasion of Ukraine. In a statement made in April 2022, then Polish Prime Minister Mateusz Morawiecki said that "in the case of Belarus, they are migrants used as a tool by a dictator. In the case of Ukraine, they are refugees fleeing the horrors of war" ("Morawiecki: Z Białorusi…"). In other statements, the Polish PM has tended to use words including "brothers," "neighbors," and "guests" to describe those arriving from Ukraine, with media statements in general steering toward this depoliticized language (Pietrusińska). On the other hand, the government and the Border Guard tend to describe those crossing through the Białowieża Forest as "illegal migrants," with right-wing commentators going further by using words such as "invaders" [*nachodźcy*] (Chołodowski)—echoing the language used around the spruce bark beetle "invasion." As Achille Mbembe put it: "The desire for an enemy, the desire for apartheid, for separation and enclosure, the phantasy (*sic*) of extermination, today all haunt the space of this enchanted zone. In a number of cases, a wall is enough to express it" (Mbembe 23). The border wall is thus rendered as a physical manifestation of racial capitalism, and those on the inside of the enclosure are using all available means to protect it.

This terminology of invasion is just one in a lexicon of dehumanization. Dehumanization similarly comes up in the notion of hunting, present in several of the texts published in the Researchers on the Border report, including this one by Joanna Sarnecka:

> The image of the hunt is clearly indicative of the dehumanisation of the hunted, which (…) has been used as a tool of propaganda and to arouse fears of people from other countries. On the one hand, hunting is an activity associated with a position of power, of domination, of holding the right over dehumanised people, who—as such—can be harmed. On the other hand (…) uniformed services—representatives of the social order and the "human" law—pursue people hiding in nature. (Sarnecka 98, own translation)

This dehumanization of people on the move raises crucial biopolitical questions: who gets to be supported by the state, and who is left to die? In *Feminist Anti-Fascism*, feminist scholar Ewa Majewska writes of the Polish state's bio- and necro-political policies with regard to the Polish feminist movement, her thoughts on the systems built by the PiS government can be translated toward the humanitarian crisis as well. The questions can be asked of all those present in the broader ecosystem of the Białowieża Forest. Sarnecka argues that the Forest is "an important actor in the drama, a participant, a victim, and our partner" (Sarnecka 101, own translation)—it can protect from the sight of authorities, or its difficult terrain can be the cause of injury or death. She writes that the Forest has become a site of death not just of the natural cycles of more-than-human passing and rot, but also of the human, by the hands of uniformed people and their border policies. People on the move thus become treated as invasive species, which the state authorities intend to remove—the "disturbance"—in order to protect the integrity of the ecosystem.

To understand how a reading of the logging conflict in the Białowieża Forest can inform readings of the humanitarian crisis it is insightful to take a closer look at a particular State Forests project. From 2017 to 2020, DNA samples were collected from the Białowieża National Park by State Forests to create a DNA bank of threatened and endangered species. The project was completed in January 2021 and 105 species from the Białowieża forest have entered the Kostrzyca Forest Gene Bank as part of the Białowieża DNA barcoding project. On its website, State Forests described the need for a DNA archive of the Białowieża forest as follows:

> The project's goal is to find fragments of tissues and seeds of the most precious plants within the Białowieża Forest area. (…) Most of the species selected for collection is [*sic*] currently endangered, therefore it is absolutely essential to secure their genetic resources in Kostrzyca Forest Gene Bank. Experts claim that because of the expansive spruce stands' dieback within the Białowieża Forest, it was the last chance to collect seeds and parts of plants from certain habitats. (State Forests, Kostrzyca n.p.)

This framing of the spruce dieback caused by the spruce bark beetle as the reason for the cryo-preservation of the forest's biodiversity is the crucial link for unraveling what we propose as *caring extraction* in the Białowieża logging conflict. Which human and other-than-human actors are presented as *disturbances* is important for understanding divergent projects of valuing life in the Forest. Indeed, State Forests stated elsewhere that "the forest is a work of man" (Środowiska qtd. in Żuk and Żuk 66) pointing to its anthropocentric notion of forest governance. Often this human mastery over the natural world is described as care for "vulnerable" forests. The notion of *caring extraction* resonates with Carrie Hamilton and Yasmin Gunaratnam's hesitations around environmental ethics of care from a feminist perspective:

> Care also carries with it the risk of turning the other into an object, a victim, and the carer into the rescuer. An urge to "save" trees may frame other-than-human vegetation as "passive participants" in their own destruction or salvation, rather than as active forces of environmental change. (Hamilton and Gunaratnam 3)

This "saving" also echoes Thom van Dooren's "regimes of violent care" (92). Van Dooren describes how conservation priorities for specific species can be to the detriment of less valued species. Importantly, this can also be applied to discourses around human migration, where some human lives are valued less than others as the discussed treatment of people on the move from Ukraine versus other areas shows.

Resistances

In the wake of state authorities intervening in the Białowieża Forest both during the logging conflict and during the current humanitarian crisis, the Forest has become a site of fierce resistance and protests. From 2016 to 2018, protestors and activists demonstrated in sections of the Białowieża Forest in response to logging activities. State Forests referred to the Forest Protection Act and described its response to protestors, who were criminally prosecuted until the EU ruling, as follows: "People blocking work in the forest and staying in prohibited areas violate the law, so it is the duty of forest guards to intervene" (State Forests, "Frequently Asked Questions"). Żuk and Żuk argue in an analysis of the political discourses during the protests that "the persecution of ecologists organising peaceful grassroots protests were a case in point that limiting civil rights and a crisis of democracy threaten prospects for environmental protection" (Żuk and Żuk 14). This is a clear example of the emergency framework and its application against human and other-than-human actors. Żuk and Żuk suggest that government intervention through logging feeds off a discourse of protecting local heritage and livelihoods. At the same time, it frames ecological conservation discourses and the efforts of activists to preserve the integrity of the forest as "cosmopolitan" and "detrimental to national interests" (Żuk and Żuk 68). The state position here is both undemocratic and anti-ecological and driven by an economic perspective of the Forest that seeks to make the Forest vulnerable, frames the protestors as a threat, and yet simultaneously extracts from the Forest.

Importantly, some of the dynamics of the treatment of activists in the Forest and the making of zones of emergency have resurfaced in the humanitarian crisis. The number of people that have attempted to cross the Białowieża Forest, dubbed by some people on the move as *the Polish jungle*, with their final destination often wishing to be Germany or other Western European countries, is not known. The numbers known are the Polish Border Guard's daily reports on the number of prevented attempted crossings, often also including the numbers of arrests they have made of "couriers," which is the word they use for the drivers involved in transporting people on the move between destinations, and the number of people the activist network Grupa Granica know to have helped since August 2021. No reported number is a complete representation of the scale of the situation. The physical violence with which people on the move are met by the Border Guards on each side of the border has been widely documented by activist groups. Beatings, destruction of phones, documents, and other equipment, pushbacks, or what Grupa Granica calls *round-ups and expulsions* are a daily occurrence and many people become trapped between the two sides. This is on top of the difficult natural conditions of the Forest, especially in the winter months, and the psychological violence of powerlessness from NGOs and other organizations, as well as fear and the criminalization of help by state authorities.

This denial of care as expressed by the state authorities is met directly with gatherings of care from grassroots networks. The role activists play on the ground in the Białowieża Forest is focused mostly on immediate humanitarian help—providing people on the move with water, fresh clothes, SIM cards, powerbanks, food and warm drinks, and providing first aid. While this has now changed and a small number of more established humanitarian organizations, particularly those dealing with medical aid, have begun working with Grupa Granica, during the pandemic period when the borderland was declared a state of emergency zone, not even medical workers were able to enter and thus much of the responsibility was placed on local residents of Podlasie.

Katarzyna Potoniec writes of Podlasie as a "transit place," similar to many others on migration routes that become "waiting rooms" or "stopovers" on the journeys made by people on the move. This transition into a place of transit forces local residents to take a side. Those who joined the humanitarian effort turned their garages into storage spaces for essential equipment, organized safe transportation, learned to use compasses and thermovision equipment, learned to hide from authorities, learned how to treat wounds—and passed on the knowledge they can to those they meet in the Forest "despite the constant threat of officers patrolling the area in helicopters" (No Borders Team "Solidarność"). They would stop to talk, offering a moment of connection, and drink a cup of tea together. However, these meetings last as little time as possible in order not to attract the attention of the authorities. Care can also be given in other forms, even in death. Several of those who have lost their lives on the border have been buried in the Muslim cemetery of the local Tatar minority in Podlasie. The complexity of the region, which is home to many minorities, is crucial in understanding the ecosystem of care, and its lack, around this crisis.[5]

As Potoniec argues, what also forms are networks, connections, and friendships between the local residents, and activists in other places in the country, and abroad—networks of care and solidarity in the face of violent and exclusionary border politics, brought together by shared values and convictions. This form of activism could be described as a form of "weak resistance", a term proposed by Ewa Majewska. Weak resistance is a feminist alternative to the masculine and heroic displays of agency. It invites alliances, heterogeneity, and non-individualism. It is neither glamorous nor individualistic, but functions on the premise of building a new reality based on a utopic idea of the common(s), and not on the reclaiming of the current system. It is about building a sense of belonging through the finding of "common ground" (Majewska 82). The criminalization of providing aid to people on the move on this route means that these actions are not seen as glamorous and heroic in the mainstream. The networks, which form around the activism on this border brought together through political conviction, can thus be seen as utopic in their creation of an alternative commons.

Conclusion: Toward More-Than-Human Solidarity

We have analyzed the Białowieża Forest as a border ecology where divergent imaginaries of invasion, vulnerability, and othering meet and overlap and ecological relationality escapes the containment of nation states. The conflicts surrounding the Białowieża Forest have shown a nationalist politicization of more-than-human entanglements in the Anthropocene, of who gets to care and what care means here. Problematic value judgments about which species are preserved and which humans are granted protection are grounded in questions of natural resource management, national enclosures, and capitalism. The complexity of Białowieża's social ecologies and its history of occupation have shown the limits of territorial and national sovereignty; as Mick Smith argues "the problem for national sovereignty seems to be the inescapable territorial permeability of causes and effects in an ecologically interconnected world" (Smith 196). These developments show the ideologically motivated violence in charged nationalist spheres of border ecologies. We argue for understanding nation states as naturecultural practices that mobilize selective notions of national "nature," "culture," and belonging, often through the creation of zones of exception or emergency such as during the Covid-19 pandemic.

As written by the No Borders Team, the Białowieża Forest "is reluctant to accept visitors, but at the same time gives them shelter" (No Borders Team "End"), playing a double role depending on how those who move through the border region interact with it, in collaborative moments forming a constellation of *more-than-human solidarities*. Bringing together our analyses of the logging conflict following the spruce bark beetle outbreak and the deadly humanitarian crisis along the EU border that runs through the forest, we build on Petra Tschakert's proposed concept of *more-than-human solidarity* for thinking across borders and species lines. What can be seen in the protest movements that emerged in the forest since 2016 are collaborative and imaginative approaches and, in both cases, more-than-humans and the Białowieża Forest itself play the role of equal participants.

It is important to remember the people and other animals who have been injured and lost their lives in the brutal border fortification. Throughout the text, we have argued that both situations are the result of similar exclusionary processes, led by national(istic) agendas expressed in the mechanisms of the State Forest authority and the Border Guard. While careful not to draw an equation mark between them, we have highlighted the entanglement of the environmental and humanitarian crises within this border ecology. We have argued that the agenda of governing bodies has been to deny the agency of more-than-humans to decide on processes of recovery and repair, as well as to impose a demarcation of category, labeling some as less-than-human through language and actions that indicate who is entitled to care and protection, and who is not. In both cases, the new border wall that cuts through the Białowieża Forest, standing between the manufactured borders of Poland and Belarus, becomes a physical manifestation of the promise to "protect" that in fact is a rendering of the denial of care.

Notes

1 We use the term "humanitarian crisis" throughout the text following the current guidance of Grupa Granica, which is suggested in response to the government using the term "migrant crisis." In addition, we use the term "people on the move" instead of "refugees" or "migrants" to encompass the broadest group of people coming through the Polish-Belarusian border; these terms are used in charged ways in the media and we discuss this briefly later in the text.

2 The Law and Justice party was in a majority government between 2015-2023. Its politics were populist, nationalist, and anti-environmental, and have resulted in multiple stand-offs with the European Union. The election in October 2023 brought a new coalition government led by the center-right Koalicja Obywatelska (Civic Coalition) led by new Prime Minister and former head of the European Council, Donald Tusk. Though there have been some policy and narrative changes around other topics, border politics between the two governemnts are very similar.

3 At the time of writing in July 2023, Facebook posts from Grupa Granica report this number to have risen to 48.

4 In addition, as we were finishing this text, PiS announced a draft legislation for a referendum on the relocation of refugees from other EU countries to Poland to take place on the same day as the 2023 general election, with the aim of using the highly charged opinions on the humanitarian crisis as a method to mobilise voters. The referendum was unsuccessful due to civic mobilisation to refuse voting cards, which meant the threshold for a binding referendum was not reached.

5 Hajnówka, a town near the Białowieża Forest, is, for instance, home to large Belarusian and Ukrainian speaking Orthodox community, which forms a minority in the heavily Roman Catholic rest of the country (Blavascunas 19) and speaks to the ethnic diversity of the border region.

Works Cited

Badaczki i Badacze na Granicy. "Działania interwencyjne: raport z realizacji projektu". *Ośrodek Badań nad Migracjami Fundacja*. Active Citizens Fund, 2022.

Blavascunas, Eunice. *Foresters, Borders, and Bark Beetles: The Future of Europe's Last Primeval Forest*. Indiana University Press, 2020.

Chołodowski, Maciej. "Kryzys na granicy polsko-białoruskiej. Skandaliczne wezwanie Ziemkiewicza do 'spuszczania wp*****lu nachodźcom'". Gazeta Wyborcza, 16 April 2023, https://bialystok.wyborcza.pl/bialystok/7,35241,29666733,kryzys-na-granicy-polsko-bialoruskiej-skandaliczne-wezwanie.html. Accessed 26 June 2023.

Chomiuk, Marcin. "Ochojska: Skuteczność muru na granicy jest zerowa". *Dziennik Gazeta Prawna*, 20 November 2022, https://www.gazetaprawna.pl/wiadomosci/kraj/artykuly/8591371,zapora-mur-na-granicy-polsko-bialoruskiej-straz-graniczna-janina-ochojska.html. Accessed 26 June 2023

van Dooren, Thom. *Flight Ways: Life and Loss at the Edge of Extinction*. Columbia University Press, 2014.

Fiałkowska, Kamila. "Obsesja systematyzacji—czyli co o migracjach sądzi Straż Graniczna, ustami swojego generała na wykładzie inauguracyjnym". *Badacze i Badaczki na Granicy*, https://www.bbng.org/obsesja-systematyzacji. Accessed 21 June 2023.

"Frontex Executive Director Visits Poland's Border with Belarus". *Frontex*, 4 October 2021, https://frontex.europa.eu/media-centre/news/news-release/frontex-executive-director-visits-poland-s-border-with-belarus-LAS4dG. Accessed 26 June 2023.

"Główne zadania". *Frontex*. https://frontex.europa.eu/pl/nasza-dzialalnosc/glowne-zadania/. Accessed 26 June 2023.

Gómez-Barris, Macarena. *The Extractive Zone: Social Ecologies and Decolonial Perspectives*. Duke University Press, 2017.

Grupa Granica. "Cykliczny raport Grupy Granica z sytuacji na pograniczu polsko-białoruskim. Grudzień 2022-Styczeń 2023". *Helsińska Fundacja Praw Człowieka*, https://hfhr.pl/aktualnosci/raport-dzialania-grupy-granica-zima-2023 Accessed 26 June 2023.

Hamilton, Carrie, and Yasmin Gunaratnam. "Environment." *Feminist Review*, vol. 118, no. 1, April 2018, pp. 1–6, https://doi.org/10.1057/s41305-018-0096-9. Accessed 3 July 2023.

Jurszo, Robert. "Przejścia dla zwierząt w murze przez Puszczę Białowieską. 'Kompromitacja GDOŚ i całego resortu środowiska'". *Gazeta Wyborcza*, 17 March 2022, https://wyborcza.pl/7,177851,28230671,znamy-wielkosci-przejsc-dla-zwierzat-w-murze-w-puszczy-bialowieskiej.html. Accessed 26 June 2023.

Łopienski, Marcin. "Martwy żubr na granicy polsko-białoruskiej. Zranił się o ogrodzenie? SG komentuje sprawę [ZDJĘCIA]". *eska*, 21 January 2022, https://www.eska.pl/bialystok/martwy-zubr-na-granicy-polsko-bialoruskiej-zranil-sie-w-ogrodzenie-sg-komentuje-sprawe-zdjecia-aa-rqPP-ExPd-fdmj.html

Majewska, Ewa. *Feminist Anti-Fascism: Counterpublics of the Common*. Verso Books, 2021.

Mbembe, Achille. *Critique of Black Reason*. Duke University Press, 2017.

——— "The Society of Enmity". *Radical Philosophy* 200, November/December 2016, pp. 23–35.

Mikusiński, Grzegorz, Jakub Witold Bubnicki, Marcin Churski, Dorota Czeszczewik, Wiesław Walankiewicz, and Dries P.J. Kuijper. "Is the Impact of Loggings in the Last Primeval Lowland Forest in Europe Underestimated? The Conservation Issues of Białowieża Forest." *Biological Conservation*, vol. 227, 2018, pp. 266–74. https://doi.org/10.1016/j.biocon.2018.09.001.

"Morawiecki: Z Białorusi chcą przedostać się migranci, z Ukrainy przybywają uchodźcy". *Dziennik Gazeta Prawna*, 11 April 2022, https://www.gazetaprawna.pl/wiadomosci/kraj/artykuly/8399583,morawiecki-z-bialorusi-chca-przedostac-sie-migranci-z-ukrainy-przybywaja-uchodzcy.html. Accessed 26 June 2023

No Borders Team. "End of September at the Border". *nobordersteam.noblogs.org*, 1 October 2022, https://nobordersteam.noblogs.org/2022/10/end-of-september-at-the-border/. Accessed 28 June 2023

———. "Solidarność w dobie wojny i przesiedleń. Anarchiści i Anarchistki przeciwstawiają się wykorzystywaniu uchodźców jako broni na granicy Polski i Białorusi". *nobordersteam.noblogs.org*, 1 April 2022, https://nobordersteam.noblogs.org/2022/04/solidarnosc-w-dobie-wojny-i-przesiedlen-anarchisci-i-anarchistki-przeciwstawiaja-sie-wykorzystywaniu-uchodzcow-jako-broni-na-granicy-polski-i-bialorusi/ Accessed 28 June 2023.

Pietrusińska, Marta Jadwiga. "Od uchodźców do nielegalnych migrantów—dyskurs wokół ludzi na granicy". *Badacze i Badaczki na Granicy*, https://www.bbng.org/od-uchodzcow-do-nielegalnych-migrantow. Accessed 21 June 2023.

@pis_org. "My nie skazujemy ludzi na śmierć…" *TikTok*, https://www.tiktok.com/@pis_org/video/7219209429990919450. Accessed 28 June 2023.

Potoniec, Katarzyna. "Podlasie jak miejsce tranzytowe". *Badacze i Badaczki na Granicy*, https://www.bbng.org/podlasie-jako-miejsce-tranzytowe. Accessed 21 June 2023.

Rexhepi, Piro. *White Enclosures: Racial Capitalism & Coloniality along the Balkan Route*. Duke University Press, 2023.

Rozporządzenie Ministra Spraw Wewnętrznych i Administracji z dnia 20 sierpnia 2021 r. (Poz. 1536). Dziennik Ustaw Rzeczpospolitej Polskiej, 20 August 2021, https://isap.sejm.gov.pl/isap.nsf/download.xsp/WDU20210001536/O/D20211536.pdf. Accessed 20 June 2023.

Samojlik, Tomasz. "Traditional Utilisation of Bialowieza Primeval Forest (Poland) in the Period from the Fifteenth to the Eighteenth Centuries". *The End of Tradition*, edited by Ian D. Rotherham et al., Wildtrack Publishing, 2010, pp. 150–65.

Sarnecka, Joanna. "Las patrzy na mnie ludzkimi oczami". *Badacze i Badaczki na Granicy*, https://www.bbng.org/las-patrzy-na-mnie. Accessed 21 June 2023.

Smith, Mick. *Against Ecological Sovereignty: Ethics, Biopolitics, and Saving the Natural World*. University of Minnesota Press, 2011.

State Forests. "Frequently Asked Questions about the Białowieża Forest, Lasy Państwowe", 2017, https://www.lasy.gov.pl/en/information/news/frequently-asked-questions-about-the-bialowieza-forest. Accessed 3 July 2023.

———. "Kostrzyca Forest Gene Bank Collects DNA of Endangered Plants in the Białowieża Forest, Lasy Państwowe", 2018, https://www.lasy.gov.pl/en/information/news/kostrzyca-forest-gene-bank-collects-dna-of-endangered-plants-in-the-bialowieza-forest. Accessed 3 July 2023.

———. "Spruce Bark Beetle in the Bialowieza Primeval Forest", 2016, https://www.lasy.gov.pl/en/information/all-about-bialowieza-forest/spruce-bark-beetle-in-the-bialowieza-primeval-forest. Accessed 3 July 2023.

Subramaniam, Banu. "The Aliens Have Landed! Reflections on the Rhetoric of Biological Invasions." *Meridians*, vol. 2, no. 1, 2001, pp. 26–40. www.jstor.org/stable/40338794. Accessed 3 July 2023.

Tomczak, Małgorzata. "Syryjczyk zmarł w szpitalu w Białymstoku. Spadł z muru na granicy polsko-białoruskiej". *Oko.press*, 24 April 2023, https://oko.press/upadek-z-muru-smierc-na-granicy-polsko-bialoruskiej. Accessed 26 June 2023.

Tschakert, Petra. "More-Than-Human Solidarity and Multispecies Justice in the Climate Crisis". *Environmental Politics*, vol. 31, no. 2, 2022, pp. 277–96.

Tsing, Anna. *The Mushroom at the End of the World: On the Possibility of Life in Capitalist Ruins*. Princeton University Press, 2015.

Żuk, Piotr, and Paweł Żuk. "Between Private Property, Authoritarian State and Democracy: Clearing Trees in Cities and Destroying the Białowieża Forest in Poland." *Capitalism Nature Socialism*, 1 December 2020, pp. 1–21, https://doi.org/10.1080/10455752.2020.1849335. Accessed 3 July 2023.

Part 4

Creative Resistance and Utopian Glimmers

12 "Preservation is an action, not a state"

DIY Utopian Enclaves and Ways out of Post-Pandemic Surveillance Capitalism in Sarah Pinsker's *A Song for a New Day*

Jari Käkelä

Sarah Pinsker's debut novel, *A Song for a New Day* (2019), was published just before COVID-19 became our real-world pandemic. Pinsker's prescient novel imagines the aftermath of a near-future period of pandemics and terrorism that leads to lockdowns, congregation bans, and social distancing, creating a world where life moves online via commercial applications. While I will explore how this setting evokes dystopian futures of surveillance capitalism, I also discuss how her novel presents Do-It-Yourself (DIY) communities as enclaves that foster rejuvenating action, inspired by the emotional power of music. Pinsker explores these themes through the characters of Rosemary Laws, a young artist recruiter for StageHoloLive (SHL), a company that distributes virtual live music in the post-pandemic world, and Luce Cannon, a rock star whose rising career was halted by the lockdowns. Pinsker's novel, I argue, presents the lockdown world as one ruled by surveillance capitalism, and imagines local communities as utopian enclaves that can resist such exploitation and direct their rejuvenating energy outward. Examining these themes, I discuss how Pinsker gestures toward the environmental effects of online existence, and, focusing on music and communal live experience, juxtaposes face-to-face experience with isolation in the virtual world.

From Surveillance/Platform Capitalism to Utopian Enclaves and Platform Socialism

While Pinsker's novel has yet to receive extensive critical attention, it has been noted as a fruitful "philosophical testing ground" for "analyses of what we are willing to give up to feel safe" (Morgan 5). Imagining this lockdown safety through online interactions on digital platforms that monopolize people's lives, Pinsker's setting aligns with ideas put forth by Shoshana Zuboff in *The Age of Surveillance Capitalism* (2019) and Nick Srnicek in *Platform Capitalism* (2017).

Through the concept of *surveillance capitalism*, Zuboff examines how businesses in the digital economy shift from making products and providing services to monetizing personal information. While this follows old capitalist laws of "competitive production, profit maximization, productivity, and growth," surveillance capitalism also exploits the "behavioral surplus" of user data accumulated

DOI: 10.4324/9781003345770-17

by digital platforms such as Facebook to predict and manipulate consumer behavior (Zuboff 66–67, 96–97). People using the services are often unaware or indifferent to how their data is exploited, although it can lead to the erosion of democracy when used for political purposes, as in the 2016 U.S. presidential elections (Zuboff 377, 278). Surveillance capitalism overlaps with Srnicek's concept of *platform capitalism*, digital infrastructures enabling interactions between user groups but also relying on data extraction and network effects to generate profit for the platform providers (24–26). Srnicek sees the tendency toward monopolization and constant user engagement as exacerbating the data accumulation on various *platforms:* Google, Instagram, and Spotify extract user data for advertising and product services, but especially Amazon arches over several platform types, from e-commerce to logistics, on-demand marketing, the gig economy, and cloud-based services (24–26). As such movement accelerates data exploitation, Srnicek proposes countermeasures of creating public platforms "owned and controlled by the people" (47, 57).

Zuboff and Srnicek have been criticized for not providing concrete solutions to the issues they highlight (DiBella; Koh), but in *Platform Socialism* (2022), James Muldoon considers practical ways to counter platform capitalism. Muldoon suggests reorganizing digital economy around "social ownership of digital assets and democratic control over the infrastructure and systems that govern our digital lives," charting strategies like platform cooperatives, participatory governance and delivering services at local level to subvert the power dynamics of platform capitalism (3, 85). While Muldoon offers pathways for reclaiming agency through community-focused initiatives, his solutions face the practical difficulty of raising community consciousness among platform users to implement such strategies (Forestal 3). Pinsker, however, presents the power of music as a way to inspire activism and foster communal connections.

Pinsker seems to imagine such activism through *utopian enclaves*. While the utopian tradition in science fiction is well established—already Darko Suvin called utopia its "socio-political subgenre" (38)—especially Fredric Jameson has examined the problems of sociopolitical uniformity and stasis of traditional utopias. Jameson sees speculative fiction as able to challenge stagnant uniformity and explore what he calls utopian desire, the collective yearning for utopia (84). This aligns with critical utopias that reject fixed blueprints and imagine directions toward which to strive, exploring imperfections within utopias to fruitfully engage with the Blochian "hope principle" (Moylan 10; Rogan 314; Freedman 69). Jameson sees attempts to realize these utopian longings as often taking place in "an imaginary enclave within real social space," possible only through "the momentary formation of a … pocket of stasis within the … forces of social change" (15). These utopian enclaves can "offer a space in which new wish images of the social can be elaborated and experimented on" free of pressures from the outside society, but their momentary nature also means the surrounding society eventually overruns them, even as it may subsume parts of their models for something better (Jameson 16–17). Although Pinsker's utopian enclaves share this momentary nature, her work also recognizes that utopian activism should be directed outward.

World Slipping Away: Lockdowns and Surveillance of Bodies and Data

Pinsker portrays the events leading to lockdowns and extractive surveillance capitalism as they halt Luce's rising career and shape Rosemary's childhood memories. Reflecting American 9/11 traumas, the novel begins with an unexplained wave of terrorist attacks, leading to "ominous requests from the government to get home, and stay home" (45, 62). These attacks are followed by an unexplained pandemic that is seen only in confused glimpses, such as an ER full of patients "flushed, sweating, shivering, moaning" (90–91), as the pandemic, similarly to COVID-19, overwhelms the healthcare system. In brief asides, Pinsker explores both the immediate and the long-term effects of the lockdowns. At the height of the turmoil, people can only helplessly read headlines calling "for people to stay home, for health and safety," but the situation gradually becomes a numbing "something something legislation something" and a sense of the "world slipping away" with "the promises of temporary change becoming less and less temporary" (91, 114). The lockdowns also exacerbate social disparities as people who cannot afford to leave the city are blamed for "working and attending school when they should have stayed home" (114), not unlike how COVID-19 exacerbated social inequalities (Perry et al.). Longer-term effects are implied by Rosemary's observations ten years later as she travels a landscape of "walled compound[s], the houses of those who had fled the city with more money than her family, or a trailer enclave, for those with less" (Pinsker 79). Rosemary's perspective also conveys the effect of growing up online: talking to someone in a bar, she cannot tell "if the grin was a leer" because virtual avatars are "so much easier to read than actual human faces" (122), depicting a world where a generation is overwhelmed by crowds and in-person interaction because nearly all social interaction is now conducted virtually.

As emergency lockdowns become permanent, many jobs disappear and society opts for convenience and a feeling of safety, leaving tech companies to provide societal functions in the virtual world of "hoodspace" (34) that runs on the logic of Amazon home deliveries, Netflix subscriptions, and Spotify streaming services. This also enforces class differences, seen in Rosemary's disdain for "her lousy old Hoodie" (124), which is incompatible with services available via more expensive models. Technology heightens class distinctions also on the societal level. Initially, Rosemary works at the online customer service of Superwally, a giant conglomerate company employing rudimentary virtual reality that leaves her uneasy of how "Superwally avatars weren't fancy enough to show much emotion" (59), but this changes as she is recruited to SHL and gets a state-of-the-art Hoodie to enter "high-end hoodspace" (34).

In addition to enforcing socioeconomic hierarchies, hoodspace is a force of homogenization. Superwally avatars are "set to age thirty-three … to project the right mix of experience and youthful enthusiasm" (16). Furthermore, work performance is standardized by constant surveillance present throughout Rosemary's workdays that start with a Quality Control call where both sides are monitored to "stick to the script" (15). She is also aware Superwally may "track her after she clocked out" (20), which recalls Foucault's notion of the panopticon in *Discipline and Punish* (1977), as applied to workplace surveillance where constant awareness of surveillance regulates behavior, like at an Amazon warehouse (Manokha 547). Studies of this panopticon dynamic at workplaces highlight how difficult it is for individual workers to resist such surveillance, leading "the managerially imposed control system to be accepted along with all other terms and conditions" (Sewell and Wilkinson, qtd. in Bain and Taylor 4). Rosemary's work performance is further monitored by gamification that makes it like collecting achievements in *Pokémon Go*:

> She forced a smile and answered.
>
> "Good morning. You've reached Vendor Services.... How may I help you?" *Good job! Your customer can hear your smile!* scrolled at the corner of her eye....
>
> The call disconnected. Rosemary glanced at her reward center. She had lost her bonus point for problem-solving efficiency—the call had gone two minutes over optimum—but got another one for refusing a gift. She was 157 points away from a merit raise. (Pinsker 17–19)

Overall, Rosemary's work is rigidly controlled to minimize variation and increase efficiency. Spending her workdays "enclosed in her Hoodie" leaves her feeling like "there were no real people, just voices and messages and lines of code" that she feeds with "data and packages and money" (20–21), hinting at the alienated isolation of work and life in surveilled virtual reality. This sense of entrapment in surveilled grunt work reflects the new American working-class reality (Ajunwa et al.), amplified by Rosemary being "told since high school" that a Superwally job is the best she can get, making leaving "unthinkable" (Pinsker 59). This is further emphasized by the "mandatory inspirational poster" in Rosemary's room that doubles as her remote office, linking the working conditions to Orwellian Big Brother type of control (see Crick 146). The poster reads: "Listen. Learn. Communicate. Our goals are speed and efficiency. Hang on! Don't hang up! You are valued but replaceable" (Pinsker 14), underlining the strain of such control on individuals.

Flocks of Drones: The Effects of Online Consumption

Luce's observations provide a view into the effects of online consumption when, ten years after the lockdowns began, she goes on the road to play in-person shows at secret venues that resist the still-enforced congregation ban. In Luce's view, the world in perpetual lockdown leaves "no jobs for the humans, other than consumption … itself a full-time occupation," and while she sees the "reasonableness" of trading "jobs as makers for … consuming from the comfort of our homes" because of the freedom it seems to provide, she views succumbing to convenience as self-entrapment (261). Luce's perspective also hints at the ecological implications of lockdown consumerism.[1] As Pinsker's focus is urban and anthropocentric, environmental implications come through hinted absences rather than direct descriptions. People are heavily drawn to hoodspace which enables some online sociability but isolates them from the outside world that is populated by self-driving cars and drones delivering online shopping (260), to the extent that they seem to replace or be mistaken as natural environment. This comes across as Luce encounters a massive Superwally distribution center, circled by what she had "assumed were starlings or sparrows [but] were in fact drones, rising in a stream, a flock, a cloud, to head to points unknown" (260–61). Also, as Rosemary views birds through augmented reality in hoodspace, spotting "migrating flocks, flying north, tagged on the wing with the BirdGoggles app," it awards her a "'First of the Season!' birding badge" for sighting what amounts to a disruption of buntings' migration patterns (79, 121). Overall, drones seem a more common sight than birds, and Rosemary "witness[ing] a hawk attacking a drone" (121) hints at the real-world drone effects on bird behavior (Rebolo-Ifrán et al.).

Ubiquitous drone delivery also implies massive energy consumption as "the whizzing flocks of package and surveillance drones" are everywhere (144), and there are no signs of any reduction in the amount of bought and delivered goods. Energy consumption of digital economy has been a concern in many real-world studies, especially with the crypto market (Huynh et al.) and drone deliveries (Koiwanit), and merely manufacturing the number of drones in Pinsker's novel would be a significant environmental concern. The need for resource conservation is mentioned in passing as Rosemary notices how "even fancy hotels weren't immune to conservation laws" and "used a gray-water system like the one at the farm" (143–44), but the novel never further explores these implications of environmental scarcity.

At the same time, for most characters the world seems rather strictly restricted to the inside, where people are in hoodspace, while the outside is somewhat run-down and occupied by drones and self-driving trucks. At the start of the novel, Rosemary is mostly preoccupied with gaining better access to the hoodspace because it seems to offer more freedom than her rural home, which to her feels like "a wind farm in nowhereville" (103) and a sign of isolation amid the abandoned infrastructure of her parents' past; "shopping centers turned detention centers turned Superwally distribution centers" (78–79). Although these asides imply the acceleration of consumerism and its environmental effects under lockdowns, Pinsker focuses more on human connection as a way to push through surveillance capitalism. The impersonal drones, as they serve endless consumption and help people stay inside their homes, are, rather, metaphorically showing that the seeming freedom of life and consumption online is confinement rather than liberation.

Their Profile of You: Extraction and Exploitation through Tracking Data and Bodies

In addition to literal surveillance of bodies, Pinsker's novel depicts algorithmic tracking that is key to surveillance capitalism. When Rosemary enters the higher-resolution hoodspace, its augmented reality view is laced with advertising where "ads for the latest Patent Medicine song and Nightlights birch beer hovered in the cumulus clouds" (79). These targeted advertisements are now, in Zuboff's terms, exploiting the behavioral surplus generated from Rosemary's use of the SHL platform and selling her more of what she just consumed. Initially, Rosemary is not bothered by this exploitation of her consumer habits, instead, similarly to Superwally workplace surveillance, she seems to merely accept it, thinking: "why care that ads tracked your interests if you had nothing to hide?" (231). This parallels Zuboff's notion of the "psychic numbing that inures us to the realities of being tracked, parsed, mined, and modified" by surveillance capitalism where "I have nothing to hide" is a defense mechanism of choosing ignorance (11).

The "congregation laws," and "curfews" put in place during the years of terrorism and the pandemic—measures similar to ones taken during the COVID-19 pandemic—are still enforced ten years after all turmoil has subsided. Pinsker never explains these laws, and they become merely a mundane part of the post-pandemic life, accepted for "feeling safe" as "there hadn't been any bombs or major outbreaks since they'd been enacted" (231). The laws also further solidify the monopolistic position of the surviving music provider, StageHoloLive (230). Although the political situation is only implied, it seems to combine elements of classical dystopias—where nonconformists stay underground—with elements of surveillance capitalism's "dispossession cycle" that creates complacency about companies' anti-democratic extraction of user data (Zuboff 138–39). The people Rosemary meets in alternative communities point out the restrictiveness of a society where data tracking is ubiquitous and explain how businesses like Superwally and SHL have a "vested interest in keeping you scared," funding anyone who will "keep the congregation laws in place, the curfews, anything that keeps people inside and using their products" (230). Compared to the real-world aggressive dispossession cycle of surveillance capitalism and the expansionistic drive of monopolistic platforms, this seems only a slight extrapolation.

Pinsker parallels the homogenizing surveillance at Rosemary's Superwally job with the experience of a rigidly planned virtual SHL concert that "closes down the conversation between performer and audience" (102). To produce an experience of seeing the show as a member of the audience, each SHL artist performs in an isolation booth separated from the rest of the band, resulting in a similar sense of entrapment as Rosemary's earlier Superwally job. At the same time, the "enormous space" of the SHL production hangar, "filled with small modular rooms" (83), provides a metaphor for a world under surveillance capitalism: Since "unpredictable behavior is the equivalent of lost revenue" (Zuboff 154), even when people could be together, they are isolated and the platforms where they conduct much of their lives create predictable behavior to maximize the revenue extracted from them. This aim of maximizing predictability is apparent also as Rosemary finds Luce's illegal live music club in Baltimore. While it is part of her job to find new talent by going to events that skirt the congregation laws, this entails correctly feeding the algorithms by considering which artists have "that mainstream appeal" of SHL material, skipping ones who have "the charisma, the presence, the musical chops" but are "too political" (Pinsker 201). Artists have always felt pressures to conform to mainstream tastes, but this is further exacerbated as SHL and Superwally are now the only legal ways to distribute music. Pinsker emphasizes the notion of feeding the algorithms as Joni, another musician Rosemary meets at Luce's club, explains, after Rosemary has searched for a song she heard in a restaurant: "Even if you're paying for ad-free, they're adding to their profile of you, waiting for a moment to sell it back to you in some way, or sell you to somebody else" (230). This aligns with Zuboff's surveillance capitalism as "the puppet master" that extracts profit from predicting and affecting our behavior (376).

Exploiting the congregation laws, SHL is also effectively culling independent music scenes and eliminating competition. Rosemary discovers this when she unwittingly betrays the trust of Luce's community and police raid her club for violating congregation laws. The raid is caused by tracking that Rosemary has so far accepted; she forgets to "scrub the location" from videos she sends to SHL of bands playing at Luce's club, and the company immediately exploits this (Pinsker 251). Data tracking employed to extract revenue from Rosemary's behavioral surplus is thus turned into the surveillance of bodies. During Rosemary's furious call to SHL after the raid, they explain it as "standard protocol" to tip off the police (243). The companies' monopolistic drive is thus extrapolated to actively destroying venues "so they can't compete with you, and the audiences are forced to see their favorite bands on SHL instead of in person, because you've taken that option away," as Rosemary now realizes (243). Doing so, SHL relies on the disposability of employees and artists: while only half of new recruiters come back after their first assignment, those who stay "realize their outrage is temporary but quitting is permanent," so they "get on with their job" (243), further paralleling the "resigned cynicism" that Zuboff sees as resulting from surveillance capitalism (11).

An Attempt to Build Something Better: Utopian Enclaves in Local Communities

In the novel, resistance to this hopeless vista comes from independent art and music communities that function as pockets of real freedom instead of the illusion of freedom provided by algorithm-steered online consumption. People who refuse to be part of a corporately determined existence have gone "noncomm" to avoid online tracking. As Joni explains, it is "not anticonsumerism… but we don't want our purchases tracked, and we don't think we always need to be in contact and trackable ourselves" (Pinsker 230). By opting out, they are reclaiming agency from algorithms and building existences alternative to those under hegemonic surveillance capitalism. Although many noncomm people lean toward abolishing platforms altogether, their grassroots approaches also align with Muldoon's ideas of a more local level in *Platform Socialism*. Rosemary eventually discovers that by hacking her Hoodie, she can enter smaller platforms than SHL or Superwally, and that searching "the free web" is still an option (230–31), even if the hegemony of monopolistic platforms keeps most people inside their ecosystems.

Luce's underground club has functioned as a utopian enclave where live music still exists despite congregation bans. She somewhat jokingly calls it "either a shrine to rock as it was or an attempt to build something better" (155), but with a sincere aim "to make it a home base for every musical weirdo looking for community outside of hoodspace" (360). This makes the in-person show inclusive differently from the homogenizing hoodspace, as Rosemary realizes when she observes how the "audience demographics varied more than she'd expected: black and brown and white, teenagers and seniors and all ages in between," while in virtual shows, "most of the avs had been young and white and had fit into the five basic av body types" (159). The sense of community at Luce's club is emphasized when people instantly help Rosemary as she gets panic attacks during her initial experiences with crowds (165, 182).

In addition to Luce's enclave for music, the novel explores her Baltimore community as a locally driven alternative to a corporately defined reality. Rosemary notices differences between the surveilled downtown where her hotel boasts "individually reinforced and blast-guarded" floors and rooms "sanitized between visits" (143) and Luce's neighborhood that is full of communal life with "rowhouses … replaced with garden plots … window boxes with flowers" and a "vendor leading a pony cart full of apples and oranges" (148). As Joni explains, lockdowns enabled their area to flourish because "rents went back to manageable when all the rich people left" and municipal resources could be "reallocated more fairly,"[2] leading to "less desperation" and working toward establishing basic income (228). Showcasing the effect this has on intellectual fulfillment, Joni takes Rosemary to art galleries and bookstores with "speakers and discussion panels" on "economy, the future, books, politics, art" so that people can hear them also offline (231). Rosemary is also introduced to a neighborhood "community garden" and they discuss restoration of contaminated city soil (228), emphasizing the contrast with the sterile built-up city where people stay inside sanitized rooms and hoodspace. These glimpses display work for local-level solutions, akin to the "Utopia of the garden city" that also Jameson notes in "the various anarchist cooperatives … and the rural communes … predicated on an idea of utopian closure" (20). Joni is a representation of such localist drive, "work[ing] on making the world out here a better place for when people come back to it" (Pinsker 231), but at the same time, she seems a guardian of a closed utopia, an enclosure Rosemary later understands should be opened. These glimpses of a fictional Baltimore community echo the Baltimore School of Urban Ecology that focuses on the patch dynamics of urban social-ecological relations, helping to understand the heterogeneity of urban ecological systems by examining how smaller patches within the larger systems contribute to complexity and "interdependent problems requiring interdependent solutions" (Grove et al. 40–42, 10). Overall, Joni's neighborhood is reclaiming a local community on a more sustainable scale.

Like Luce's club, this community becomes a utopian enclave away from commercially driven online platforms, an existence new to Rosemary who marvels at the "tiny ethnic grocery stores, coffee shops, restaurants, hair salons, all small enough to skirt the congregation laws" (231). Although this enclave rises from a need to survive the aggressively commercialized mainstream, its focus on building a community decoupled from the capitalist logic of endless growth aligns Pinsker with hopepunk visions (Romano). As a further representation of more fulfilling local communities outside commercialized virtual space, Joni's accommodation resembles real-world places like squats in the Netherlands or American punk scenes and intentional communities. The kinship within these communities, as Rosemary is struck by Joni's dinner gathering between unrelated "people who treat each other like family" (233), signals that safe, creatively-inspiring existence outside surveilled virtual reality is possible.[3] This highlights the contrast between these communities, enclosed to keep them safe from surveillance, and the surveilled mainstream that is enclosed in a way that isolates and subjects people to the ills of surveillance capitalism.

A Call to Action: The Power of Music and the Move from Utopian Enclaves Outward

> This was another thing altogether.... The kick drum rose up through her bones; the bass mimicked her pulse, or her pulse mimicked the bass.
>
> Rosemary found herself moving forward with the crowd, pressing toward the stage, dancing, dancing with real people, in real life.... (182)

While Pinsker presents the utopian enclave as resistance to surveillance capitalism, her protagonists also acknowledge the issues with remaining enclosed. Jameson detected in traditional utopias the danger of stasis and in utopian enclaves a tendency for withdrawals to be too transitory to become transformative (20). Traveling the country, Luce has witnessed "endless creativity when it came to carving out space for music" (Pinsker 330), but this creative energy seems wasted when used for complex safety measures to keep insular venues from getting raided, instead of developing their art and making it more available. The fragmented insularity Luce witnesses on this tour shows her that all "communities needed new blood or they stagnated" (308), and to challenge the endless lockdown, she sets out to "outpace the change, be the change" (258) and spread the "joy sweeping through a crowd. The good contagion" (113). Juxtaposed with the now-ingrained fear of medical contagions, Luce's aim highlights her reliance on the power of live music. After discovering SHL's culling tactics, also Rosemary heads outward, seeking to subvert the system by turning recruitment into a "controlled burn" to avoid its destructive after-effects (315). While the local utopian enclaves and Luce's desire to avoid all platforms seem like attempts to return to the past, Rosemary recognizes the potential of the digital world for transformative outward action. Frustrated during her tour of small secret venues after her underground club was closed, Luce shouts to drone cameras outside Graceland, exhorting people to "sculpt yourself out of a new medium … construct ways to belong" (273–74). Posted online, this call for freedom goes viral and leads to countless people "posting responses, saying what they're going to do" to answer Luce's call (335). When Rosemary discovers the viral video, she is inspired to become "a conduit for Luce's message" (325), eventually convincing Luce to join forces with her.

Although the novel's conclusion enshrines Luce as a prophet figure of sorts, Rosemary's generation also understand they "need people who will call legislators … run for office … write articles" and connect people "out there … already working on it" (343). What the younger generation in Pinsker's novel is doing seems to align with Cory Doctorow's manifesto on resisting surveillance capitalism. Rather than "returning to a world without tech," for Doctorow the solution is techno-socialism where "it's up to us to seize the means of computation, putting that electronic nervous system under democratic, accountable control." Such a drive toward communal ownership of data is further developed in Muldoon's notion of platform socialism. Pinsker's local utopian enclaves resist the power of Big Tech, which Muldoon considers the first step in reclaiming agency from the platforms. The end of the novel also points to what Muldoon sees as next steps: to restrict the corporate reach and "foster alternative systems and processes of collaborative production" (9).

The final act of Pinsker's novel presents live music as fostering collaborative communality both online and in real life. Rosemary convinces SHL to broadcast an on-location in-person show on Luce's terms, but seizing the means of computation, she also hacks their system to give free access to the show. SHL's approach to setting up a show by a pre-lockdown legend like Luce is "like they'd discovered an intact dinosaur fossil" (351) to be preserved as playing just the old hits, but Luce counters this by Bob-Dylanesque quips such as "preservation is an action, not a state" (350), highlighting her urgency to use the show to counter the stasis brought about by exploitative capitalism. The novel's conclusion thus focuses on human connection, enabled by the emotional charge of live music. During the show, Luce talks to the audience about reversing the artificially maintained lockdown, saying "we need to take community back ourselves–nobody's going to give it to us" and "people are a risk worth taking,"[4] to steer out of the isolation maintained by surveillance capitalism and connect the pockets of action witnessed throughout the novel (360). After her speech, the band launches into a song Luce had been writing since the lockdowns began, "an instruction manual, a guide, a call to action" (361). The novel's titular song for a new day, then, is Luce's manifesto for interaction unrestricted by platforms that keep people inside and consuming. Overall, the binary in the novel between consumerism indoors and creativity outdoors is linked to setting corporate and grassroots activity against each other. While the binary is not directly equivalent to our world, through the act of literally stepping outside hegemonic system—because of how its enclosures suffocate creativity and diversity—the binary becomes an allegorical representation of hegemonic capitalism.

Various rhetorical studies point to music as a form of activism and resistance (e.g., Sellnow and Sellnow). Pinsker also continues the way science fiction has utilized "the charismatic nature of rock music—and its power to create emotions so strong that they can be read by those who feel them as transcendent" (Davies et al.). In the end, Luce becomes a sort of rock prophet to guide the next generation into the future. During the ecstatic culmination of the show, Rosemary realizes that only the transcendent shared experience where "everyone here is marked by their presence" matters (365), and in a hopeful sign, the audience includes both in-person show veterans and people "who had ventured out of hoodspace for the first time" (364). Rosemary observes the transcendent effect even in online discussions with

> people trying to figure out what to do to answer Luce's call. A law student offering to start a group to take on congregation laws, someone else saying they wanted to host a show in their basement, someone else talking about running for office on a pro-congregation platform". (366)

This shared drive—working toward a community consciousness that Muldoon's platform socialism requires—aims for change through a multitude of solutions, some through resistance by setting up new enclaves and others by seeking to move out of enclaves altogether, but all aiming for peaceful change rather than violent revolutions.

Coda: Room for Us to Exist

Earlier in the story, frustrated after Luce's club is raided, Rosemary walks on the SHL compound "in the opposite direction of manicured walking trails, into the unmapped areas" to a "cordoned-off bridge" over a river that is "nothing more than dried mud" (267). The dried-up wilderness outside SHL's hyper-manicured park parallels the struggling ecosystem of music in hidden-away clubs, setting Rosemary on her mission of subverting the exploitative system. In the novel's coda, Luce and Rosemary exit the venue after their first success at that subversion. They walk toward another "overgrown path … to a tiny footbridge" (367), but this stream is "fast moving" and they see how "an owl darted out of the darkness to skim the water. It came away with a small silver fish writhing in its talons, then disappeared back into the woods" (368). At this moment, the novel's metaphor goes full circle: the artificial world of surveillance capitalism may dry up the organic stream of music, but it can be rejuvenated through inspiring communality like Luce's live show, if such more organic patches are allowed to exist as well. As Rosemary aims to change SHL from the inside, Luce comes to see that "there's something in helping them see there's room for us to exist" (368). This suggests that when the drive for action is organic, it can inspire communities both online and in person, and platforms matter less than how they are used. Thus, Rosemary's action becomes one of countering surveillance/platform capitalism not by abolishing technology, but by moving toward platform socialism and harnessing technology to work for the people, not against them.

In relation to the dystopian fiction of the past few decades, Pinsker sounds a more hopeful note and imagines a tangible call to action that begins at the local level. While it is set amid the after-effects of a pandemic, Pinsker's novel is far from such dramatically ruined futures as the post-apocalyptic, disease-ridden biopunk worlds of Paolo Bacigalupi's *The Windup Girl* (2009) or Larissa Lai's *The Tiger Flu* (2018). Rather, Pinsker extends beyond warnings of future disasters to explore grassroots solutions that can spur larger-scale developments. By singing her way out of dystopia, Pinsker writes a hopeful ode to the power of DIY music communities as utopian enclaves that can use the "good contagion" of musical experience to inspire change in the world at large.

Notes

1 While it was enthusiastically reported during early COVID-19 lockdowns, for example, that the perpetual smog in New Delhi had cleared when people stayed home (Kinver), these moments seem to have been mere fleeting glimpses, not signs of nature rebounding from pollution (Gautam et al.).

2 Such hopeful post-disaster visions align with works like Kim Stanley Robinson's *2312* (2012), which imagines an optimistic Venice-type future for NYC after sea level rise, resulting in "the city … improved by the flood" (100).
3 This communality likens Pinsker's novel to works such as Becky Chambers' *The Long Way to a Small, Angry Planet* (2014), where inclusion in terms of sexual identity, gender, race, and different species are unquestioned parts of much of the world.
4 In contrast to the legitimate real-world health concerns, many still have over gathering in large groups, in the novel in-person meetings are safe and lockdowns are artificially maintained for companies to preserve their market share.

Works Cited

Ajunwa, Ifeoma, et al. "Limitless worker surveillance." *California Law Review*, vol. 105, no. 3, 2017, pp. 735–776.

Bain, Peter, and Phil Taylor. "Entrapped by the 'electronic panopticon'? Worker resistance in the call centre." *New Technology, Work and Employment*, vol. 15, no. 1, 2000, pp. 2–18.

Crick, Bernard. "Nineteen eighty-four: Context and controversy." *The Cambridge Companion to George Orwell, Cambridge.* Edited by John Rodden. Cambridge UP, 2007, pp. 146–159.

Davies, Hugh, et al. "Music." *The Encyclopedia of Science Fiction.* Orbit. https://sf-encyclopedia.com/entry/music. Accessed 20 June 2023.

DiBella, Sam. Review of *The Age of Surveillance Capitalism*, by Shoshana Zuboff. *LSE Review of Books*. 4 November 2019, https://blogs.lse.ac.uk/lsereviewofbooks/2019/11/04/book-review-the-age-of-surveillance-capitalism-the-fight-for-the-future-at-the-new-frontier-of-power-by-shoshana-zuboff/. Accessed 20 June 2023.

Doctorow, Cory. "How to Destroy Surveillance Capitalism." *One Zero*, 26 August 2020, https://onezero.medium.com/how-to-destroy-surveillance-capitalism-8135e6744d59. Accessed 20 June 2023.

Forestal, Jennifer. Review of *Platform Socialism*, by James Muldoon. *Contemporary Political Theory*, 29 July 2022, https://doi.org/10.1057/s41296-022-00576-3. Accessed 20 June 2023.

Freedman, Carl. *Critical Theory and Science Fiction*. Wesleyan UP, 2000.

Gautam, Alok Sagar, et al. "Temporary reduction in air pollution due to anthropogenic activity switch-off during COVID-19 lockdown in northern parts of India." *Environment, Development and Sustainability*, no. 23, pp. 8774–8797. https://doi.org/10.1007/s10668-020-00994-6. Accessed 20 June 2023.

Grove, Morgan J., et al. *The Baltimore School of Ecology: Space, Scale, and Time for the Study of Cities*. Yale University Press, 2015.

Huynh, Anh Ngoc Quang, et al. "Energy consumption and Bitcoin market." *Asia-Pacific Financial Markets*, vol. 29, no. 1, 2022, pp. 79–93.

Jameson, Fredric. *Archaeologies of the Future. The Desire Called Utopia and Other Science Fiction*. Verso, 2007.

Kinver, Mark. "Then and now: Pandemic clears the air." *BBC News*, 1 June 2021, https://www.bbc.com/news/science-environment-57149747. Accessed 20 June 2023.

Koh, Sin Yee. Review of *Platform Capitalism* by Nick Srnicek. 5 June 2017, https://blogs.lse.ac.uk/lsereviewofbooks/2017/06/05/book-review-platform-capitalism-by-nick-srnicek/. Accessed 20 June 2023.

Koiwanit, Jarotwan. "Analysis of environmental impacts of drone delivery on an online shopping system." *Advances in Climate Change Research*, vol. 9, no. 3, 2018, pp. 201–207.

Manokha, I. "The implications of digital employee monitoring and people analytics for power relations in the workplace." *Surveillance and Society*, vol. 18, no. 4, 2020, pp. 540–554.

Morgan, Glyn. "New ways: the pandemics of science fiction." *Interface Focus*, vol. 11, no. 6, 2021, pp. 1–7.

Moylan, Tom. *Demand the Impossible: Science Fiction and the Utopian Imagination*. Methuen, 1986.

Muldoon, James. *Platform Socialism: How to Reclaim Our Digital Future from Big Tech*. Pluto Press, 2022.

Perry, Brea L., et al. "Pandemic precarity: COVID-19 is exposing and exacerbating inequalities in the American heartland." *Proceedings of the National Academy of Sciences*, vol. 118, no. 8, 2021, pp.1–6.

Pinsker, Sarah. *A Song for a New Day*. Berkley, 2019.

Rebolo-Ifrán, Natalia, et al. "Drones as a threat to wildlife: YouTube complements science in providing evidence about their effect." *Environmental Conservation*, vol. 46, no. 3, 2019, pp. 205–210.

Robinson, Kim Stanley. *2312*. Orbit, 2012.

Rogan, Alcena Madeline Davis. "Utopian Studies." *The Routledge Companion to Science Fiction*. Edited by Mark Bould et al. Routledge, 2011, pp. 308–316.

Romano, Aja, 'In the Era of Trump and Apocalyptic Change, Hopepunk Is Weaponizing Optimism', *Vox*, 27 December 2018, https://www.vox.com/2018/12/27/18137571/what-is-hopepunk-noblebright-grimdark. Accessed 20 June 2023.

Sellnow, Deanna, and Timothy Sellnow. "'The Illusion of Life'" rhetorical perspective: An integrated approach to the study of music as communication." *Critical Studies in Media Communication*, vol. 18, no. 4, 2001, pp. 395–415.

Srnicek, Nick. *Platform Capitalism*. Polity Press, 2017.

Suvin, Darko. *Metamorphoses of Science Fiction: On the Poetics and History of a Literary Genre*. Yale UP, 1979.

Zuboff, Shoshana. *The Age of Surveillance Capitalism: The Fight for a Human Future at the New Frontier of Power*. Profile Books, 2019.

13 Pandemic Dramaturgy

Co-Designing the Performance *Dying Together/Futures* with COVID-19

Alice Breemen

Introduction: Finding a New Middle Ground in Theater

When, in 2020, the COVID-19 pandemic hit the world, theater practitioners were forced to invent alternative modes of performing and, consequently, had to explore different modes of spectatorship. In many countries, theaters were temporarily shut down to limit the spreading of the virus. Gathering as a live audience in a theater building was prohibited due to the risks of spreading the virus, which caused theater makers to come up with alternative practices, like streaming theater or using public spaces as a stage. Convening with theater makers and audience members in one place no longer determined the theatrical event; instead, a myriad of new dramaturgies that dealt with the pandemic reality appeared.

One of these performances with an unconventional dramaturgy was *Dying Together/Futures*, by Dutch theater company Building Conversation, presented in between lockdowns in September 2020 in the Frascati theater in Amsterdam, The Netherlands. Even before the COVID-19 pandemic, Building Conversation was already researching alternative theatrical formats, which were mostly conversations inspired by dialogical practices from all around the world. These include, for example, Indigenous conversational practices without words and dialogue practices invented by the Jesuits (Aers et al. 36). *Dying Together/Futures* was the new addition to their series entitled *Dying Together* that engaged with various events in which people, more-than-human entities, systems, and ideas died simultaneously.

The *Dying Together* series started in 2018 with *Dying Together/Humans* and focused on specific situations in which humans collectively died, like the terrorist attack on the French Bataclan concert hall, or the crash of the airplane Germanwings 9525, and the capsizing of a boat at Lampedusa Island. The second part *Dying Together/Earth* was created in 2019 and dealt with the death of plants, animals, or the relationship between all beings on earth. Situations that were explored—both singular mass death events as well as "slower" forms of death and violence—were the mad cow disease 20 years ago in Great Britain, the Amazon forest exploitation, and the suicide of a Japanese businessman in the Aoikagara forest (Building Conversation). Both iterations not only researched the various situations of dying together but also what this might mean for living together.

DOI: 10.4324/9781003345770-18

Building Conversation explored these events in the physical presence of the audiences and described their performances as "embodied thought exercises," in which participants entangled and disentangled actors, perspectives, and situations related to dying together. Instigated by the performers, participants in the created event represented elements of these situations in a constellation, a theatrical term borrowed from astronomy that points to stars that together form a figure when imaginary lines are drawn between them. As such, participants had to give up the traditionally passive role of audiences, and instead become actors responsible for the establishment of the performance. They were thus clearly part of the construction and continuation of the performance.

Being confronted with the new reality of the pandemic, the theater makers created the third part in 2020 that engaged with multiple perspectives involved in the worldwide spreading of the COVID-19 virus. This part was called *Dying Together/Futures*, because at this particular moment in time not only did people die at the same moment but the pandemic also highlighted collective futures, ideologies, and lifestyles that became uncertain. This third part engaged the audience in developments related to climate change and its connection to "future moments of dying together" (Building Conversation).

This chapter explores how *Dying Together/Futures*, by playing with theater traditions and expectations, presents a rehearsal space for participating in newly formed (future) assemblages and alliances that do not end when the performance is over.[1] The central argument is that, in the dramaturgy, the more-than-human enters and seeps through the invisible holes left unattended, claiming its own place on the stage (Haraway). I analyze how the participants experience the performance as a constant movement of shifting between perspectives and roles and, at the same time, engage with the virus as an unpredictable, more-than-human actor in the dramaturgy. The concept "onto-epistemological theatricality" is used to expand on how this double process works in the dramaturgy. This leads to considering how the performance might be a middle ground (Braidotti)—a space where participants can practice and play with relational entanglements that characterize the Anthropocene.

Part 1. A Milky Way of Concepts

Creative Concepting

This analysis of *Dying Together/Futures* will be performed by practicing creative concepting (Van der Tuin and Verhoeff). Creative concepting is the practice of working with concepts in the field of the creative humanities as described by Van der Tuin and Verhoeff in their book *Critical Concepts for the Creative Humanities* (2022). Following Mieke Bal, the scholars use concepts as "mini-theories" (Van der Tuin and Verhoeff 6) that help understanding phenomena but also incorporate the position of the researcher. Each analysis offers a "situated reading." This means that the analysis is connected to a certain time and place, as well as the (background of the) researcher. This approach connects past, present, and future while acknowledging the relational entanglements of both researchers and their objects of study: "situated at the intersection of critical and creative practices, concepts for the creative humanities are reflexive of the contemporary moment yet future-oriented in thinking with and towards possibilities for emergence, change, and transformation" (Van der Tuin and Verhoeff 2). The creative humanities transcend the boundaries of academia, the claim being that "[s]cholarship is already-also taking place elsewhere in an intellectualized and intellectualizing domain of design, art, and culture—and how scholarly work is never isolated from, but always already part of culture" (4). The creative humanities are thus a discipline that is operating at crossroads, switching back and forth between various timescales and between academia, the arts, and the world at large.

The practice of working with concepts is inspired by three theories complementing each other at different levels. These theories all emphasize the importance of "situated knowledges" cf. Haraway (qtd. in Van der Tuin and Verhoeff 4). First, there is a reference to Raymond Williams' *Keywords: A Vocabulary of Culture and Society* (1976), placing the book in the context of other "dictionaries" and collections of concepts. Important is Williams' idea that concepts "[a]re expressive of the entanglements of ontology, epistemology, ethics, and aesthetics of a particular period in time and history" (Van der Tuin and Verhoeff 5). The second theory is Mieke Bal's notion of travelling concepts, in which interdisciplinarity is at the center of cultural analysis. Van der Tuin and Verhoeff note that this does not only apply to academic disciplines but also incorporates art and design in working with and creating concepts. This last point then situates creative concepting in the realm of a third theoretical position of new materialism, relating to "knowing in being" cf. Barad (qtd. in Van der Tuin and Verhoeff 6).

Onto-Epistemological Theatricality

Working with concepts in the ways described above invites criticality and "[e]ntails an ethics that takes responsibility for each position involved" (Van der Tuin and Verhoeff 7). I, therefore, engage with an onto-epistemological approach of theatricality, meaning that I focus on how the performance balances between being in the middle of the movements of the earth (being), and I try to gain insights into how the construction of knowledge is going on in the performance (knowing). Theatricality as a concept has had a different meaning in different places and ages (Davis and Postlewait). With regard to the epoch that is called the Anthropocene,[2] I am interested in how, as a concept, theatricality at work in an epistemological sense questions worldviews and shakes up systems of thought by confronting a spectator with how the mechanisms of a theatrical production function (Röttger; Féral and Bermingham; Schramm). From an ontological point of view, spectators can experience theatrical events in tune with how the earth rocks and flows (Lavery; Weber). An onto-epistemological approach of theatricality relates to the Anthropocene in the sense that humankind finds itself at once in the middle of a new geological epoch and is at the same time in search of new ways of moving in it, of coming to terms with the alterations to the planet caused by a small part of humanity, whereas, in some places, humans (and more-than-humans) only feel the consequences of global warming, which can be experienced in theatrical events simultaneously.

Dramaturgy

Dramaturgy is central in this chapter in two ways—both in regard to the organization of an embodied thought experiment in the performance *Dying Together/Futures* and as a concept for analyzing relational entanglements the Anthropocene confronts us with, in this case, with the COVID-19 virus. In 1994, dramaturge Marianne van Kerkhoven argued in the State of the Union that, in the current world, the "major dramaturgy should be the most important." The major dramaturgy is the world outside the theater and the exchange that takes place between what happens in the theater and in the world at large. Van Kerkhoven writes that this is not only the space of the theater building but also the city surrounding the theater building, and the world around it—even reaching to the sky and the stars. The boundaries between these spheres are not closed. Van Kerkhoven states that they are "made of skin," they "are made of pores, they breathe" (7). The exchange between the theater production and the world at large is in constant motion.

Adding to the notion of a major dramaturgy, Augusto Corrieri asks what kinds of dramaturgies are suitable to the challenges presented by the Anthropocene. Amongst others, Corrieri includes the on-stage appearance of "non-human subjects" who "have gatecrashed the party" (236). He claims that it is not possible to make only theater on the small, human scale; instead, it is required to set aside the focus on theater's "'here and now' into an infinity of sites and extended temporalities" (236). Stevens characterizes this description of dramaturgy of the Anthropocene as deep dramaturgy, relating to the notion of deep time. It is a concept that is used to mark geological or cosmic time, which is measured in billions of years (Mauelshagen). It is the timescale of the stars, the big bang, and the evolution of the Earth. For theatrical dramaturgies to engage with such timescales, it is not possible to work only with human measurement tools. Therefore, the dramaturgy becomes a tool for thinking: "As such, dramaturgical doing can be understood to work with a two-sided impetus of creative thinking: as both a creative process of experimentation and thinking and also designing the conditions for future experimentation and thinking" (Van der Tuin and Verhoeff 85).

Pandemic/Contagion/Virus

The third concept related to in this chapter is the term pandemic itself. Only since 1853, the word pandemic was used to point to the worldwide spread of a disease. Etymologically, the meaning of pandemic stems directly from Old Greek, and translates as "pertaining to all people; public, common" ("Pandemic"). The term is, as such, connected to a disease travelling across borders, locations, and timescales, and therefore operating in a complex network of involved actors. The virus as more-than-human participant in these networks—in the theater and also outside—is a force that now affects all kinds of relations. As a concept that relates to infection, or contagion, it becomes a figure to think about issues that have come up in the Anthropocene. In their entry "Contagion," Van der Tuin and Verhoeff describe this double meaning as follows:

> Contagion, or contagious spread, is inherently generative in its connecting of bodies of people, organisms, things, and ideas. This spreading productively disrespects analytical distinctions and borders such as between human, technology, and nature and traverses historical, social and theoretical categories such as gender, race, nation, and class. (Van der Tuin and Verhoeff 58)

In the notion of pandemic, the contagious spread of the virus, there is thus a dimension of being always already connected to others, human and more-than-human alike. According to Judith Butler, the virus is travelling from body to body: "What another breathes out, I can breathe in, and something of my breath can find its way into yet another person" (Butler and Yancy). She argues that the pandemic confronts us with dying and demands for sharing acts of mourning what is lost in the pandemic and in particular how that is related to issues of vulnerability (Butler and Yancy). This vulnerability is a global state of being and has become more visible during the pandemic.

In a similar argumentation, Braidotti states that the coronavirus pandemic "[h] as proved a powerful catalyst in revealing often concealed or hidden degrees of social inequality" (27). The connection to others is thus not only limited to the human species but requires, according to Braidotti, also integrating more-than-human actors in the formation of new assemblages. She argues that we therefore need a "shift of perspective" to find a new middle ground where we can do this (26).

Both Butler and Braidotti thus observe how we are in the middle of a process in which more-than-human actors are trying to claim a place centre stage but, at the same time, there are no ready-made categories for where and how this change of dramaturgy has to happen. A middle ground must be invented where it is possible to explore these relations and, as such, expand and question the very systems of knowledge production itself. In the next part of this chapter, I analyse how onto-epistemological theatricality as practice and concept highlights relational entanglements in the pandemic dramaturgy of *Dying Together/Futures*, in particular, and offers such tools to navigate in the Anthropocene—the epoch in which these entanglements are becoming more prominent.

Part 2. Dancing with the Virus[3]

The work of Building Conversation is characterized by a very specific format for each performance. In their recent publication, *Building Conversation: The Scripts* (2022), Peter Aers et al. give an overview of how most of their performances are constructed. Among others, in every performance the notion of participation is central. This means that the makers of the performance and the audience are equally important in the establishment of the work. Consequently, an outsider position is not possible. Lotte Van Den Berg describes this as following:

> When practising [sic] participation, you can only work together, be it with audiences, artists, programmers, journalists and researchers. More than that, you are invited to be present with everything you are and to let go of certain roles or responsibilities. We have to be willing to take part with the stumbling, uncertain, irritated, fearful and over-enthusiastic parts of ourselves. ("Actual Participation" 78)

My analysis is thus written from the point of view of one of the participants, in line with Van der Tuin and Verhoeff's notion of a situated reading. I bring my own background and profession of a theater scholar with me in the unfolding of the performance and, in doing so, I give an analysis from being within the performance. From this position, I reflect on what these experiences and insights brought me with regard to living in the (aftermath) of the pandemic. Therefore, in this part of the chapter, the writing style subtly changes to more informal, in order to emphasize how the dramaturgy worked in regard to being immersed in the unfolding event and actively taking part in designing the constellations.

Before the Beginning

The performance *Dying Together/Futures* had a layered dramaturgy that began even before the performance had started inside the theater and also did not end when the curtain fell—there were no curtains, and a particular part of the performance was archived and digitalized. The play with these various dimensions of space and multiple timescales caused a mingling of the spheres of the theatrical event and the outside world (Van Kerkhoven): a before or an after were not clearly defined but, instead, an entanglement of dimensions was established. Through the theme of the performance—dying together through the virus and the consequences of living together—the virus was both an imaginary, invited guest to play in the performance, as well as an actual dangerous actor in the invisible 1.5-meter space between the participants. The dramaturgy became a tool for an embodied thought experiment and, in connection with the virus, was highlighting how everything is entangled with everything else. As participant, I became involved in this game by the invitation of the performers, which was extended throughout the framed event and continued after I had left the theater.

Setting the stage is important in the work of Building Conversation, because, according to Daan't Sas, one of the theater makers of the company, this marks the moment of switching from the introduction to the part where participation is required (Sas 83). *Dying Together/Futures* began with an explanation of the "rules" of the performance. Still outside the theater space, even outside on the street before the theater, one of the performers told the audience that the obligatory 1.5-meter distance should be kept at any time between the participants. Together with a group of other participants, I walked to the empty theater space, where the chairs and curtains were removed and, thus, where no visible boundaries between participants and performers were installed.

Upon our arriving in the theater space, the makers of Building Conversation explicitly asked for our participation in the establishment of the constellations. Van Den Berg writes that "[y]ou can only say yes, if there is the possibility of saying no" ("The Possibility" 82). Confirming that we were going to participate in the event made all of us responsible for the performance and made us also part of a group with whom we were going to spend the evening.

Kati Röttger argues for theatricality as a methodology for performance analysis that foregrounds the spectator as a constitutive element of the performance. She states that "[t]he structural condition of theatre could be described in very broad terms as an event of seeing and being seen in the here and now, or the constellation of performing and spectating in one and the same moment" (Röttger). This idea foregrounds the role of the spectator and positions theatricality as a "mode of perception" (Röttger) while opening up the possibility of considering theatricality as an "epistemic category" (Röttger).

This implies that, through balancing between illusion and materiality in the here and now of theater, worldviews can be questioned. According to Röttger, a moment of instability takes place in between "truth and fiction, and it is up to the spectator to make a decision for one or the other." Theatricality, as such, has medial qualities as well, functioning as the "middle state" from which "perspectives on the world" appear through a balancing act between spectator, the medium itself, and the historical context that operate on a visible and invisible level (Röttger).

The construction of the performance *Dying Together/Futures* was explained beforehand, and the rules for how to engage in the performance presented. The dramaturgy required participation from everyone; there was no possibility to disconnect or stay out. Throughout the performance, many moments of instability appeared—for example, when I was asked to play various roles in the constellations, which made me switch back and forth between myself and others. A shift of perspective was an explicit part of the dramaturgy, which was caused by the "play" with various points of view and the assignment for the audience to literally embody other perspectives that were part of the pandemic. Furthermore, in between all participants and actors, we embodied the fact that there were relations to explore as well, which was also part of the performance. In the invisible space between the participants, the virus (or the suspicion of it) was present as well, connecting us and letting us move in a certain way, as such agent taking an active lead in the dramaturgy.

Another possibility was that the performers would not ask you to take part in the formation of a constellation. Still, this did not mean that you would be left outside, because this switch from representing something yourself to looking at the constellation only heightened the awareness that we were all in this together—both in the space of theater and in the pandemic reality at that time. Your role would be "spectator" in this case, but not passive onlooker, because you would still be connected to the others and the virus in the constellation that was unfolding in front of you.

During the Explorations of Constellations

After everyone had picked a spot in the empty theater, one of the makers asked us to read a pamphlet that explained the theme of the performance and the particular situations of dying together, related to the coronavirus pandemic, that we were going to explore that night. A performer explained that the participation of the audience was required for building the performance around the creation of three *constellations,* building up and broadening the perspective on the pandemic and the actors involved. After this introduction, we were going to create the constellations. One of the makers would sketch the situations related to the pandemic. One of them had a bunch of cards, with elements connected to the situations written on them, which they would read out loud. Then, they would ask us individually if we wanted to represent these elements in the constellation. This happened at the start of each constellation. The makers timed these constellations, and, after each constellation, they would state where we were, how many participants there were, and what time it was.

In the first constellation, which applied to our own position in the pandemic, one of the makers would guide us to a certain spot in the theater space and, as soon as everyone was placed in the constellation, they asked us to bodily relate to the other participants. In practice, this meant walking very carefully through the space, being aware of the 1.5-meter distance, and sometimes witnessing other participants forget to keep distancing. While walking around, I was looking at someone and trying to make eye contact, and trying to find my way through this uncomfortable situation we were in together.

The second constellation would broaden our perspective to various elements connected to the pandemic—such as a whale, who, for the first time in his life, swims in a clean sea when factories are closed to limit the virus spreading, a banana box that is now used for transporting corpses, a bank employee from Singapore who suddenly is home constantly and, for the first time, feels connected to her neighbors. These perspectives were mostly differing from my own perspective and location, but were now part of the pandemic, helping me in forming a more detailed picture of it.

In the third constellation, we were asked to represent ideas and concepts related to the pandemic. These were, among others, the idea of white supremacy—the idea that it matters where you are born—where the performers asked various participants to represent a baby born at a different place on earth. Also, this constellation explored the idea of some people amassing even more wealth while others—for example people working in the so called “crucial professions”—could not afford to stay at or work from home and be safe from the pandemic. The theater makers asked us questions to which we could relate with our bodies: for example, how the idea or concept that each of us represented would change in 20 years, in 40 years, in 70 years, in 200 years, and in 1,000 years. So, we were moving our bodies through these time and space scales, rehearsing for a future yet to come. While moving my body through and to these ideas and the other concepts that were represented, I was confronted with how hard it is to think about these deep timescales. The timeframes of 20 years were relatively easy to imagine, because hopefully I will be still alive in 2040. However, the other timescales were harder to think about. What will become of the planet and of humanity, and all the more-than-human actors intertwined with them, in 1,000 years from now?

Carl Lavery describes theater as a medium that "deterritorializes" and inserts a "hollow place" in human identity that connects it to an outside that Lavery calls "a non-human fold that unwittingly and oxymoronically remains in relation by parting" (165). As such, it never "arrives" and is always "parting with" (165). Within this never-arriving movement, it is possible that "theatricality suspends the metaphysical nexus that holds together accepted notions of time, space, and identity" (165). Lavery writes the following:

> Not only does theatricality … "unground" anthropocentric notions of time and space—frameworks habitually used to domesticate the earth—it can also attune audiences to a different way of existing with and on the planet. It does so by positing being as something ungraspable, caught up in an immanent process of endless transformation and splitting in and from itself. (165)

The play with timescales in *Dying Together/Futures* built up and resulted in a deep dramaturgy, with the theater as a place and also no place: a hollow place in time and space that connected the here and now with past and future. In the last parts of the performance, these connections were mixed up and expanded in the Archive of Reflections.

After and More after: Theatricality and the Timescale of the Long Durée

The last part of the performance consisted of three parts: adding to an ongoing archive that will return in future performances of the *Dying Together* series, the after talk, and the online archive. All three elements invited reflection, which is often part of Building Conversation's work, and were as much part of the performance as the embodied thought exercise with constellations. When the third constellation was finished, the makers invited the participants to contribute to the Archive of Reflections, which connected history and future human audiences to each other. An archive installation was brought to the theater space. We were asked to add an archive card with our reflection on the performance. We could also read what other audience members had written and could add our own reflections for audiences coming after us. This was extended to an online Archive of Reflections, which everyone with Internet access can read. Digitalizing the archive adds an extra layer to the dramaturgy because it has now been made accessible for people who were not at the performance in the first place.

Digital traces of the performance will remain, however, in terms of the long durée, or the timescale that is difficult to relate to in the here and now. Since it takes place over long periods of time, it remains to be seen if the archive will continue to exist. Yet what the archive confronted us with as participants of the performance is the very notion that there remain traces of us, while also connecting us to future audience members and asking us to reflect on other actors involved in the pandemic. Another participant wrote on an archive card, for example, how the performance let him/her/them reflect on how the virus and other actors are currently hard to separate from each other: "Making contact with the virus while standing in front of her. Staring into each others [sic] eyes and wanting to feel anger. Instead, we smile at each other. You are just as much part of this as I am" (Archive of Reflections).

This is an engagement with the virus via the other participant that represents the virus. On this card, the distinction between the representation of the virus and the virus as an autonomous actor faded. At that moment, the participant referred to the virus as an actor, and realized they are together in "this." The idea of being connected is reflected on by yet another participant in this entry:

> I'm sitting in the middle of the room, looking around at all of the people around me. They are already representing something or someone, while I'm only a spectator looking around, noticing the relations that everyone has with each other, trying to connect in some way. This is a mesmerizing thing to look at. To feel like a connected outsider, seeing how the others are trying to be together. (Archive of Reflections)

This participant remembered the moments when he/she/they were outside a constellation and did not (yet) take part in it. However, the participant realized that there was no such thing as being outside. The participant felt confronted with the connection of actors that was happening before his/her/their eyes and realized they were also part of the action while still being outside. This particular experience of the participant makes clear how, in the performance, and outside, everything is entangled. It is precisely at this point that the performance can be placed in a larger debate on which dynamics in current society the dramaturgy confronts the participants with, and how this affects the future.

Open Ending

Dramaturgy is movement, ecology, and an organization of elements in motion. The strict dramaturgical framework of *Dying Together/Futures* was made clear in advance and throughout the performance, which in combination with the more open parts of the performance invited onto-epistemological theatricality. Onto-epistemological theatricality is located precisely on the unstable boundaries between both and opened space for the virus to participate. It is a practice and a concept that facilitates an in-between, a hollow place, or a middle ground. While being in the performance, participants shared the responsibility for what emerged. Being asked to represent something is inherently theatrical and, at the same time, confronted us with our responsibility toward the various positions that are part of the pandemic. Who am I to represent someone on the other side of the world? The one moment, one was representing oneself; the next moment, one was attempting to represent the virus and perhaps looking the actual virus in the eye. What happened in between switching these positions had the potential of unsettling us and asked us to critically reflect on our own role in this rhizomatic constellation of actors.

Stevens states that the role of theater in the Anthropocene (in general) and dramaturgy (in particular) is finding connections between the human and the "natural" worlds, a process that becomes "[l]ess painful as we accept and even embrace our human vulnerability and similarity to the non-human and work towards a non-human-centric restoration of the environment that our species has so profoundly altered and damaged" (96). The dramaturgy of *Dying Together/Futures* instead engages with an assemblage of all kinds of perspectives and goes against the idea that there even is such a thing as "our species," resulting in reflections on the multitude of possible positions from which people experience crises related to climate change, such as the pandemic. These perspectives constantly shifted from a personal point of view in the pandemic to other participants and situations, and then to concepts and ideas of being together in the pandemic and its aftermath. In the dramaturgy were thus multiple scales and temporalities integrated, all of which depended on the position from which one was looking.

Dramaturgy can, as such, become a tool that, at the same time, confronts us with our own role in visualizing the process of dying related to climate change, along with other ecological cataclysms like the sixth mass extinction. In a certain way, dramaturgy can soothe this process by placing the emphasis on humankind as just one actor in a network of many. Seen this way, the concept of dramaturgy extends the theater and can be made useful for thinking about how to live and die in the epoch of climate change and alterations to the planet.

In the case of *Dying Together/Futures*, the dramaturgy is a space where we can get to know the virus because we have to literally perform *with it*, where exclusion is not possible because the virus does not discriminate. Therefore, the pandemic dramaturgy connects actors across boundaries, and offers a practice space for living and dying together in the Anthropocene, as such rehearsing for a future in which relational entanglements claim a role center stage. In the face of a present and a future where COVID-19 might be the latest of many pandemics to come, practicing how to live with the pandemic—and thus to die—might become a survival strategy for a future in the Anthropocene.

Notes

1 In this chapter, I analyze the version of *Dying Together/Futures* from 2020. This "try-out" was made under strict governmental rules and inbetween lockdowns, meaning that participants should always keep the 1.5-meter distance rule in mind and could thus not bodily engage with each other. Later versions differ dramaturgically from this try-out, because participants now were able to bodily connect with each other during the performance.

2 The term "Anthropocene" is at the centre of many contemporary debates about the name itself, or when the geological epoch started, or whose Anthropocene it is (Crutzen; Haraway; Zalasiewicz et al., etc.). In this chapter, however, I am, in particular, interested in how—with Lavery—the term "Anthropocene" can be considered as theatrical in itself.

3 In the Netherlands, one of the campaigns to motivate people to get a vaccine was called "Dansen met Janssen," which translates as "dancing with Janssen." One could get the Janssen vaccine and then go out again to party. However, the vaccine did not prevent the fact that many of these events where people came together to dance became super-spreading events, showing the unpredictability of the virus and the necessity of moving with it carefully, which became only clear after the damage was done already ("Jongeren").

Works Cited

Aers, Peter, et al, editors. *Building Conversation: The Scripts.* Building Conversation, 2022.

Archive of Reflections. Building Conversation, 2020, www.archiveofreflections.nl. Accessed 7 June 2023.

Bal, Mieke. *Travelling Concepts in the Humanities: A Rough Guide.* U of Toronto P, 2002.

———. "Working with Concepts." *European Journal of English Studies*, vol. 13, no. 1, 2009, pp. 13–23.

Barad, Karen. *Meeting the Universe Halfway: Quantum Physics and the Entanglement of Matter and Meaning.* Duke UP, 2007.

Braidotti, Rosi. "'We' May Be in *This* Together, but We Are Not All Human and We Are Not One and the Same." *Ecocene: Cappadocia Journal of Environmental Humanities*, vol. 1, no. 1, 2021, pp. 26–31.

Building Conversation. Stichting Building Conversation, 2023, www.buildingconversation.nl/aanbod-item/dying-together-toekomstbeeld/. Accessed 7 June 2023.

Butler, Judith and George Yancy. "Interview: Mourning Is a Political Act Amid the Pandemic and Its Disparities (Republication)." *Bioethical Inquiry,* vol. 17, 2020, pp. 483–87.

Corrieri, Augusto. "The Rock, the Butterfly, the Moon and the Cloud: Notes on Dramaturgy in an Ecological Age." *The Practice of Dramaturgy: Working on Actions in Performance*, edited by Konstantina Georgelou, Efrosini Protopapa, and Danae Theodoridou, Valiz, 2017, pp. 239–51.

Crutzen, Paul J. "Geology of Mankind." *Nature*, vol. 415, 2002, p. 23.

Davis, Tracy C., and Thomas Postlewait, editors. *Theatricality.* Cambridge UP, 2003.

Féral, Josette, and Ronald P. Bermingham. "Theatricality: The Specificity of Theatrical Language." *SubStance*, vol. 31, no. 2–3, 2002, pp. 94–108.

Haraway, Donna J. *Staying with the Trouble: Making Kin in the Chthulucene.* Duke UP, 2016.

"'Jongeren die gingen dansen na Janssen kunnen coronaschade verhalen op staat'." *Nos.nl*, 7 June 2023, www.nos.nl/nieuwsuur/collectie/13850/artikel/2389642-jongeren-die-gingen-dansen-na-janssen-kunnen-coronaschade-verhalen-op-staat.

Lavery, Carl. "Theatricality and Drifting in the Anthropocene." *Nordic Theatre Studies*, vol. 2, no. 1, 2020, pp. 159–78.

Mauelshagen, Franz. *Deep Time.* Universität Wien, anthropocene.univie.ac.at/resources/deep-time/. Accessed 29 September 2021.

"Pandemic." *Online Etymology Dictionary.* 7 June 2023, www.etymonline.com/word/pandemic.

Röttger, Kati E. "Staging Night Watch: Theatricality and Scenes of Crime in Peter Greenaway's Screen Adaptation of Rembrand's Painting." *Theatricality and the Arts: Theatre, Film, Art*, edited by Andrew Quick and Richard Rushton, Edinburgh UP (forthcoming).

Sas, Daan't. "Will You Join." *Building Conversation: The Scripts*, editors Peter Aers, Lotte Van Den Berg and Sodja Lotker, Building Conversation, 2022, p. 83.

Schramm, Helmar. "Einleitung: Das offene Buch der Alchemie und die Stumme Sprache des Theaters." *Spuren der Avantgarde. Theatrum alchemicum,* De Gruyter, 2017, pp. XIII–XXXIII.

Stevens, Lara. "Anthroposcenic Performance and the Need for 'Deep Dramaturgy.'" *Performance Research. A Journal of the Performing Arts*, vol. 24, no. 8, 2019, pp. 89–97.

Van Den Berg, Lotte. "Actual Participation." *Building Conversation: The Scripts*, editors Peter Aers, Lotte Van Den Berg and Sodja Lotker, Building Conversation, 2022, p. 78.

———. "The Possibility of No." *Building Conversation: The Scripts*, editors Peter Aers, Lotte Van Den Berg and Sodja Lotker, Building Conversation, 2022, p. 82.

Van der Tuin, Iris and Nanna Verhoeff. *Critical Concepts for the Creative Humanities*. Rowman & Littlefield, 2022.

Van Kerkhoven, Marianne. *State of the Union Address*. 1994. Dramaturgy Database, dramaturgydatabase.hum.uu.nl/wp-content/uploads/sites/673/2018/09/Van-Kerkhoven-State-of-the-Union-1994-in-Etcetera.pdf. Accessed 4 October 2021.

Weber, Samuel. *Theatricality as Medium*. Fordham, 2004.

Zalasiewicz, Jan, et al. "The New World of the Anthropocene." *Environmental Science & Technology*, vol. 44, no. 7, 2010, p. 2228.

14 Vitality of Nonhuman Entities

Plagues and Pandemics as Hyperobjects in Defoe, Camus, and Pamuk

Hülya Yağcıoğlu

> The plague bacillus never dies or vanishes entirely . . . it can remain dormant for dozens of years in furniture or clothing … it waits patiently in bedrooms, cellars, trunks, handkerchiefs and old papers, … perhaps the day will come when, for the instruction or misfortune of mankind, the plague will rouse its rats and send them to die in some well-contented city. (Camus 237–38)

So reads the last sentence of Albert Camus' *The Plague* (1947), pointing out the dystopic potentialities of a plague awaiting humankind in the future. Plagues and pandemics have indeed long devastated the populations of humans and other animals in history. The Bubonic Plague or The Black Death in fourteenth-century Europe, the smallpox epidemic in the eighteenth century, and the 1918–1919 influenza pandemic—commonly known as Spanish Flu—are among some relatively recent pandemics that left millions of people dead.[1] The recent COVID-19 outbreak is also a catastrophic event that has reminded us once again of the fragility of life on Earth. It has become a "hyperobject" of our age, to use Timothy Morton's term, which "seems to force something on us, something that affects some core ideas of what it means to exist" (15). As such, even very basic life forms like bacteria and inert viruses may well be "actants" with some efficacy to make a difference and change the course of history. This chapter will examine three important representations of plagues in fiction in order to theorize and contextualize the recent pandemic: Daniel Defoe's *A Journal of the Plague Year* (1722), Albert Camus' *The Plague* (1947), and Orhan Pamuk's *Nights of Plague* (2022). The theoretical background of the chapter involves Morton's theorization of "hyperobjects" and Jane Bennett's theory of "vital materiality," as well as object-oriented ontology (OOO) and Bruno Latour's actor-network theory (ANT). This chapter will argue that, however dystopic the real or fictional pandemics examined here are, they may also come with a utopian promise: as potential tools to renew our perception of other nonhuman life forms on the planet, with whom we have long co-existed, and which preexisted before the emergence of the homo sapiens. A so-called inert substance such as a "virus" may make us reconsider how we understand the entities and environments around us and become more attuned to the vitality of the nonhuman world including animals, plants, and microorganisms.

DOI: 10.4324/9781003345770-19

Vital Materialities

When the recent coronavirus disease (COVID-19) broke out in December 2019 in Wuhan, China, the catastrophic impact of the pandemic on the global economy, trade, health, and the general well-being of individuals was unforeseeable. More than 6 million people have been recorded to have lost their lives as of January 2023, according to the World Health Organization. The pandemic also caused a significant disruption to the global economy; people lost their jobs and many of them were forced into governmentally enforced quarantines. No other event has possibly had such a devastating global effect in the last few decades.

Morton defines the recent coronavirus pandemic as "the ultimate hyperobject of our age," which drastically affects the entire world and its beings on a large scale (qtd. in Morgan). Following the trend of OOO and ecological studies, Morton's work undermines long-held assumptions about the separation between humanity and nature in line with Latour and calls for an awareness of our interdependence with other beings. He theorizes "hyperobjects" as vital nonhuman entities of vast temporal and spatial dimensions, which ultimately necessitate human beings changing the way they relate to nonhuman forms. Therefore, the COVID-19 pandemic is a "hyperobject" as, in the current age of the Anthropocene, it is no longer possible to view a pandemic as an entirely human phenomenon without acknowledging the influence of and on nonhuman species.

Microbiological life forms, such as viruses and bacteria, act as important agents among the many intricate relations between species. Bacteria are living, single-celled organisms, present in almost every environment in nature. Plagues are generally caused by a form of bacterium, *yersinia pestis*, as opposed to COVID-19's coronavirus, which is normally an inert substance. Stemming from a Latin word meaning "slimy liquid" or "poison," a virus is defined as an infectious agent of small size and simple composition that can multiply only in living cells of animals, plants, or bacteria ("Virus," *Britannica Academic*). Viruses do not show any signs of life until they get inside living cells to reproduce themselves and eventually become part of the cells ("Virus," *Key Contemporary Concepts*). Because viruses come alive with the presence of a host and are unable to survive long without it, they blur the lines between inert matter and vital life. Povinelli points out how viruses "occupy an interstitial space between life and non-life, between organic and in-organic," "blurring ontological boundaries" among species. Viruses could continue living in extracellular environments (on surfaces etc.) for some time. Viruses are always in a state of becoming because "unlike parasites, they could still be alive when existing in non-life" (Povinelli).

In the case of a coronavirus infection, we see how the human body and the virus are engaged in an intricate and dynamic multispecies relation. When a virus enters a cell, it changes its genetic information and "becomes part of the cell itself and thus part of part of the body's own system," so a virus has "a capacity to disguise itself because it has no identity of its own, and to be a parasite and to use another environment as its own, and to mutate rapidly" ("Virus," *Key Contemporary Concepts*). A virus, then, is a liminal agent that deconstructs the binaries between self and other. As soon as humans are infected with a virus, the virus begins to multiply and reproduce in the host's cells. Morton gives the example of how a virus "remixes the components of the organism it enters," and notes that "human beings are cyborgs of a kind, since we are made up of all sorts of non-human components ... the very stuff that supposedly makes us us—our DNA—contains a significant amount of genetic material from viruses" (qtd. in Blasdel). COVID-19 is a hyperobject that is "literally inside us" (Morton qtd. in Meis). The virus also blurs the binaries between the inside and outside: the virus enters our bodies, which host it to survive and multiply. Thus, it is a part of us that makes us sick; and yet, it is also "not us"—it is "the other" within us, an elusive liminal agent. The coronavirus pandemic has also engendered some highly discriminatory rhetoric based on the idea of "othering": it was called "a Chinese virus" by some and considered the epitome of an "evil other," coming from an alien land.[2] Varlık also highlights how plague has long been regarded "as a temporary 'invasion' or 'alien' presence" ("New Science" 205).[3] When it comes to the coronavirus pandemic, war terminology has often been used while mentioning a virus that supposedly is as devastating and invasive as an army; it embodies a relentless enemy, which we must wage a war against. The global efforts to recognize the scope of the virus, understand its impact, and eventually defeat it demonstrate the power struggle surrounding the coronavirus pandemic. The mechanism of "othering" was not limited to nonhuman species such as animals and viruses: other human beings have become "others" too, as everyone is a potential enemy who could spread the disease. Shunning others as potential enemies and quarantining at home to avoid contact with others were essential parts of our recent coronavirus experience.

More importantly, the COVID-19 pandemic made us realize that human beings are not the only active agents. Recent theories of OOO and ecocriticism that follow a long philosophical tradition of Louis Althusser and Michel Foucault have attempted to de-center the "human." OOO holds that the external world exists independently of human awareness, hosting mutually autonomous objects. The founder of the movement, Graham Harman, rejects the privileging of human existence over the existence of nonhuman objects. Latour's ANT has similarly described the social phenomena in terms of the interplay of human and nonhuman actors. For Latour, an actant, whether human or nonhuman, is an intervener that makes things happen; it is the decisive force catalyzing an event. He uses the term "actant" rather than "actor" to rid the word of any trace of anthropomorphism and defines it as "both humans and nonhumans; an actor is any entity that modifies another entity in a trial; of actors it can only be said that they act; their competence is deduced from their performances" (237). Evidently, basic microscopic organisms like bacteria and inert viruses may well be "actants" with some efficacy to make a difference and change the course of events. Taking into consideration the enormous number of people and other animals who died historically as a result of plagues and pandemics caused by such microorganisms (among which are the Bubonic Plague in fourteenth-century Europe and the 1918–1919 influenza pandemic, which left at least 50 million dead), not to mention the drastic socioeconomic changes they have engendered, it is clear that human life is closely entangled with other nonhuman actants. Jane Bennett blends ANT and ecocriticism and theorizes what she calls "vital materiality," which runs through and across bodies, both human and nonhuman, in her work *Vibrant Matter: A Political Ecology of Things* (2010). She suggests that agency is not only the province of humans and always emerges as the effect of configurations of human and nonhuman forces and acknowledging this might change the way we examine public events. What bacteria and viruses have in common, then, is a sort of "vital force" inherent in them. Intrinsic to human life on Earth, vibrant matter or "thing power"—to use Bennett's terms—refers to an active, energetic vitality intrinsic to things that are generally conceived as inert, such as a virus. Bennett states that human bodies are also composed of vital materials; thus, it does not make great sense to put humans at the ontological center or hierarchal apex of all species. Her aim is to cultivate a more responsible, ecologically sound politics, and to inspire a sense with which all bodies are kin in a dense network of relations (Bennett 13).

Fictions of Plague

Examining plague fictions adds new dimensions to our understanding of the boundaries of human life and our entanglements with human and nonhuman interactions. Such fictions reflect humans' moral, psychological, and behavioral responses to catastrophe. In her article "What Our Contagion Fables Are Really About," Jill Lepore argues that a plague novel is where human beings abandon their humanity and become each other's enemy, as "a plague is like a lobotomy. It cuts away the higher realms, the loftiest capacities of humanity, and leaves only the animal." These narratives then depict the descent of human beings from civilization to savagery: when human beings are preoccupied with mere "survival," they sense their animality (Lepore). As such, the traits through which they proudly set themselves apart from nonhuman animals diminish (Lepore). And yet, despite the apocalyptic sense that the world has come to an end, such narratives also signal a new beginning, or rather, a better utopian new world order. This is in line with the idea of apocalypse, which refers to "a transformation, perhaps in consciousness, by which an existing corrupt socio-ecological order is turned upside down by the astonishing irruption of new hope" (McIntosh, ix). Therefore, narratives of contagion are examples of how dystopia has the potential of turning into utopia as humans only realize the preciousness of life after the threat of extinction. These narratives often envision a new world order after the plague, portraying it as utopian and offering a promise that life will be improved now that humanity has endured.[4]

Defoe's *A Journal of the Plague Year*, one of the most significant plague narratives, is based on the bubonic plague, or the Great Plague as it is called now, which claimed nearly a hundred thousand lives in London in 1665. The novel is told in great dramatic detail and is based on the narrator H.F.'s observations and experiences during the plague. The narrator's disorderly style of narration mirrors the chaotic state of the plague. The novel indeed conveys a world that is falling apart, a world on the brink of catastrophe against which human beings are helpless: the plague "is a formidable Enemy, and is arm'd with Terrors, that every Man is not sufficiently fortified to resist" (Defoe 223–24) and it "came upon them like an armed man" (Defoe 115). Under the unpredictability of the contagion, one can see similar insights into typical human behaviors, as we have experienced during the recent coronavirus pandemic. In disbelief, panic, and anger, Londoners started panicking as the number of cases increased; some fled the city and some shut themselves at home with provisions. The government took action by imposing quarantine and sterilization measures, banning public gatherings, closing down schools, and ordering a lock-down of infected individuals in their houses. Similar to the devastating effect of the current pandemic on the global and local economies, Defoe conveys how "all Trades being stopt, Employment ceased; the Labour, and by that, the Bread of the Poor were cut off" (92). There are familiar instances regarding, for example, the transmission of the disease by touch:

> It is true, People us'd all possible Precaution, when any one bought a Joint of Meat in the Market, they would not take it of the Butchers Hand, but take if off of the Hooks themselves. On the other Hand, the Butcher would not touch the Money, but have it put into a Pot full of Vinegar which he kept for that purpose. (Defoe 75)

The novel even points out the presence of asymptomatic individuals who spread the disease without knowing it. The novel reads very much like apocalyptic fiction, depicting ominous stories of infected people in great pain who commit suicide by throwing themselves into the river or mass graves, and of the loss and separation of loved ones.

Defoe's world is clearly an anthropocentric one in which human beings are at the ontological center of all other species. The plague in the novel gives way to a new perception of how human lives are closely entangled with super/natural forces. Justus Friedrich Carl Hecker's pivotal study of the Black Death in 1832 and his followers in the nineteenth century regarded plagues and pandemics as enigmatic "forces of nature that shaped societies" (qtd. in Varlık, "Plagued Legacies"). This led to what is called "gothic epidemiology" in which pandemics are attributed great agency and described as "alien, mysterious and almost unexplainable supernatural forces on human society that have the power to lead to catastrophic events and determine people's fate" (Varlık, "Plagued Legacies"). In the novel, Londoners do not seem so much preoccupied with the natural causes of the plague and try to make sense of it by looking for a divine cause: "Good people began to think, that God was resolved to make a full End of the People in this miserable city" (Defoe 97). Distraught by the great extent and severity of the calamity, some people think it was caused by "a possession of an evil Spirit" and resort to churches and fortune tellers. In *Illness as Metaphor*, Susan Sontag writes how disease, which could be considered a part of nature, "became the synonym of whatever was 'unnatural'" (74). The plague historian Nükhet Varlık likewise highlights how Western scholarship has often treated plague "as a temporary 'invasion' or 'alien' presence and has focused heavily on the effects of this 'foreign' visitor, instead of examining plague's interaction with the natural and built environment" ("New Science" 205).

Originally a disease of rats, bubonic plague is spread to humans by fleas that must find a living host after the rats die (Goodwin xii). As such, one can see a total lack of awareness and respect for nature and other nonhuman animals with which human beings cohabit. In fact, as a precaution, the government commands citizens "to kill all the Dogs and Cats … [who] are capable of carrying the Effluvia or Infectious Steams of Bodies infected, even in their Furrs and Hair," and as a result, around 40,000 dogs, "five times as many Cats," and "a great number of Mice and Rats" were destroyed (Defoe 116). This mass massacre of nonhuman, sentient animals demonstrates how human beings, situated at the top of the hierarchical apex of beings, exploit nature and other species for survival. The novel ends with the sudden retreat of the plague and the narrator's remark: "A dreadful Plague in London was, In the Year Sixty Five, Which swept an Hundred Thousand Souls Away; yet I alive" (Defoe 236). John Carrey highlights the cathartic sense in the novel in which "[the reader] is continually placed in the position of a survivor," conferring a "comforting sense of … immortality" (qtd. in Roberts ix). Thus, the novel ends with a utopian promise of "survival," of the continuation of life after the devastating catastrophe.

Written more than two centuries after Defoe's novel, Camus' *The Plague* talks about a fictional epidemic that destroys half the population in the Algerian city of Oran. The plague starts with an alarming number of rats coming out of cellars and dying in pain by spurting blood from their muzzles. The disease spreads from rats to humans, who become living hosts to the plague-causing bacteria. The novel includes some in-depth observations on the social and psychological effects of the plague, which has a lot of parallels with the recent COVID-19 pandemic, especially regarding how humans respond to contagion. "A pestilence," wrote Camus, "does not have human dimensions, so people tell themselves that it is unreal, that it is a bad dream which will end" (30). As such, the first reaction is denial: "Pestilence is in fact very common, but we find it hard to believe in pestilence when it descends upon us" (Camus 30). Then comes panic and fear because of mounting deaths as well as "feelings of separation and exile" and "fear and rebellion" (Camus 129).

The plague is depicted as abrupt, random, and inconceivable, and yet a part of life, as highlighted by Dr. Rieux's old patient's words: "But what does that mean, the plague? It's life, that's all" (236). The plague thus is a condition of being human in its absurdity and potential of evil. "The plague is man in Camus' novel," remarks Lapore as the novel depicts the alienation and brutality of modernity, where "men will always become, again, rats." The novel was, in fact, written as an allegory of the German occupation of France during World War II. Although the novel was highly acclaimed, Camus was also criticized for deploying the metaphor of a "non-ideological and nonhuman plague" for the Nazi occupation (Dunn qtd. in Judt 246).

What makes Camus' novel intriguing is the way people in Oran behave as a response to catastrophe. The novel focuses on a collective experience of people more than individual stories as "there were no longer any individual destinies, but a collective history that was the plague, and feelings shared by all" (129). In the protagonist, Dr. Rieux, for instance, one can see the moral responsibility of an ordinary doctor trying his best to do the right thing for the public. In a sense, the catastrophe acts as a common enemy in bridging the alienation among people and bringing them together to fight for survival. However, although the plague gradually vanishes, the novel ends with a threat. The last sentence of the novel indicates that

> the plague bacillus never dies or vanishes entirely … it can remain dormant for dozens of years in furniture or clothing … it waits patiently in bedrooms, cellars, trunks, handkerchiefs and old papers, and … perhaps the day will come when, for the instruction or misfortune of mankind, the plague will rouse its rats and send them to die in some well-contented city. (Camus 237–38)

One can infer an impending catastrophe awaiting humanity, lurking and repeating itself. This potential calamity will be caused by *bacillus*, a type of bacteria, that remains "dormant" and one sees how a lifeless thing may be a quasi-agent in engendering life-changing events. In *Vibrant Matter*, Bennett gives an example from Darwin's study of worms and how he regards the worms as "small agencies" whose "accumulated effects" turn out to be quite big (96). The dormant bacteria, then, are also a part of "heterogonous assemblages in which agency … is distributed across a swarm of various and variegated vibrant materialities" (Bennett 96). I argue that the last words of Camus' novel do not necessarily suggest a dystopic future outlook but urges us to acknowledge our shared existence with nonhuman entities vis-a-vis the inevitability of plagues.

The Nobel Laureate Turkish author Pamuk's *Nights of Plague* fictionalizes the outbreak of bubonic plague on the fictional island Mingheria of the Ottoman Empire in 1900. The novel was first published in Turkish in 2021, at the climax of the COVID-19 pandemic, although the author had been working on it for a few years before the pandemic struck. At the onset of the plague, the island is depicted as "an imaginary Eden" that is so beautiful that, when they first see the island, Princess Pakize and Doctor Nuri feel "bewitched" by the "inscrutable landscape, the mountains, the silence beneath the moonlight [which] were imbued with a wondrous intensity [that] belonged 'in the pages of a fairy tale'" (19). Pamuk's creation of Mingheria is "a world so detailed, so magically full, so introverted and personal in emphasis, that it shimmers like a memory palace, as if Pamuk were conjuring up a lost city of his youth, Istanbul's exilic, more perfect alter ego" (Wood). The plague, however, is like a stain on the seemingly utopian island as there is a sharp contrast between the idyllic Mingheria and its post-pandemic, wretched state. The first reaction of the islanders to the plague is disbelief and denial as in the previously discussed two plague narratives:

> It was still possible to convince oneself that this was as bad as things were likely to get, that this outbreak too, like every outbreak before it, would eventually fade away, and that one could get through it all unscathed by hiding out in some quiet, private corner, and avoid going out for a while. (Pamuk, *Nights* 203)

The authorities of the island have some initially optimistic plague containment strategies like clerks spraying vast quantities of Lysol into the air, which all reassure the public that the epidemic is under control and yet turn out to be quite ineffective in containing the plague. The people of the island, especially the Muslim population, do not obey the quarantine rules and the plague spreads at an alarming rate, ravaging the entire island.

The plague here is a local one just like the plagues confined mostly to London in Defoe's novel and to Oran in Camus' novel. The island of Mingheria, situated between the East and the West, is depicted as a liminal place. In its isolated existence, the island is even far from the capital of the Ottoman Empire. The outbreak of a plague in Mingeria—the "other" land for the West—confirms the idea that such contagions necessarily come from "outside," from elsewhere. Pamuk (or rather, the narrator, Mina Mingher) talks about how the threat of contagion often comes from the East for the Western audience, and how, with pilgrimages to Mecca or Medina, Muslims became "the world's most prolific carriers and spreaders of infectious disease" in the nineteenth century (Pamuk, *Nights* 77). Varlık likewise highlights that modern European scholarship has long dissociated itself from plague and associated it with the Orient, "the port cities of the eastern Mediterranean," and the Ottoman Empire in particular, as "by projecting the locus of the disease somewhere outside; on the other, it fashioned the Ottoman empire as a plague-exporter, against which Europe had to protect itself" ("New Science" 204). Both Pamuk's novel and Varlık's historical accounts demonstrate how plagues have historically been unfairly Orientalized.

In *Nights of Plague*, Pamuk is more interested in the political effects of the plague than its effects on humanity. The limits of the government are tested during the chaos of the pandemic. The plague thus sets the background of drastic political changes to shape the history of the island. Wood notes that, at the end of *Nights of Plague*, the utopian promise is realized: "Pamuk is nostalgically drawn to that brief utopian moment when Mingheria broke free, when it announced its impossible independence, when the plague began to fade, when the new state was full of pure potential." It is in this apocalyptic moment that the utopic dream of nationalism, of independence for the state of Mingheria from the Ottoman Empire, takes place.

The three novels studied here evidently share many similarities with the recent pandemic, especially in terms of how the public's reaction to the contagion follows a similar cycle: denial, acceptance, governments' efforts of containment, a sense of the apocalypse, and the end of the contagion with a utopian promise. They all regard the contagion as a catastrophic event: a punishment of God in Defoe's novel, a chance and absurd event in Camus' novel, and a chaotic event that leads to drastic socio-political changes in a state in Pamuk's novel. In both historical and popular imagination, pandemics have often been regarded as "apocalyptic," "isolated exceptional outbursts that do not last for more than a few months or years" and as aberrant accidents of nature that inevitably destroy human societies (Varlık, "Plagued Legacies"). As such, they all come from an anthropocentric tradition and are solely concerned with the effects of the plagues on humans without consideration of the other nonhuman forms. As opposed to the representations of "localized" plagues, COVID-19 is a global pandemic that easily spread to the entire globe due to fast-paced global mobility and travel. As it has had a tremendous effect on all parts of the globe on a large scale, it helped us see how contagion is indeed a result of a web of connections and interactions not just among human beings but also among different species. It made us question our existence and the fragility of our lives as well as our shared co-existence with other nonhuman beings. I argue that the utopian promise of the COVID-19 pandemic should be to acknowledge that human beings are not the only actors and our existence is based on an assemblage of relationships and interactions between different agents, whether human or nonhuman.

The COVID-19 pandemic, just like plagues in history, is a perfect example of interspecies interactions through viral exchange as it is believed to be a zoonic disease. The plagues in the novels examined are also transmitted to humans by rats; that is, bacteria spread to human hosts to survive. Although pathogens breach the boundaries of species to survive, the plague is still considered a merely human event in those narratives. For instance, rats die in enormous numbers in all the novels, and cats and dogs perish in Defoe's novel. Pamuk's novel depicts an instance where nonhuman animals are treated with no respect: a shepherd dog and a horse are just used to test some biscuits poisoned with ratsbane, which results in both dying painfully (223). Many pets and innocent animals have died as a result of coronavirus as well: Denmark, for instance, was reported to have culled 15 million minks after discovering a mutated strain of COVID-19 in its mink farms ("Matching the Right"). Although we share similar vulnerabilities with nonhuman, sentient animals, neither the narratives nor the real pandemic we are witnessing acknowledge and respect our entanglements with other nonhuman life forms. The Anthropocene era situates the human species "within the wider panorama of life on the planet, forcing us to acknowledge the fact that we exist in delicate equilibrium with an overwhelmingly complex yet interconnected global ecosystem" (Gibbon et al.). Considering the profound impact that human beings have had on the Earth's ecosystems, it is imperative that we acknowledge and confront the significance of the nonhuman.

In the same way the fictions of contagion studied here come with a utopian promise—as catalyzers for change for a better social order—the COVID-19 pandemic may also give us a chance to imagine a new order after experiencing the ultimate catastrophe. Arundhati Roy regards the recent pandemic as "a portal, a gateway between one world and the next" and notes that "unlike the flow of capital, this virus seeks proliferation, not profit, and has, therefore, inadvertently, to some extent, reversed the direction of the flow." In an opinion piece in *The New York Times* in 2020, Pamuk wrote that the pandemic calls for a shared sense of humanity:

> The terror we are feeling, however, excludes imagination and individuality, and it reveals how unexpectedly similar our fragile lives and shared humanity really are. For a better world to emerge after this pandemic, we must embrace and nourish the feelings of humility and solidarity engendered by the current moment.

Therefore, despite all the pessimism and desperation surrounding the recent pandemic, Roy and Pamuk are among the writers who regard it as a possibility of revival. Conversely, I argue that, rather than just going back to "normal" and establishing human solidarity, this pandemic should give way to a more radical and utopian outcome: embracing a more-than-human perspective to acknowledge the agencies of nonhuman life forms.

In sum, this chapter has examined the recent COVID-19 pandemic through three literary representations of the plague, which all depict similar human reactions to the crisis. The plague narratives studied here demonstrate how contagion may give way to a more utopian order after experiencing ultimate catastrophe. Having had unprecedented effects on the entire globe and its inhabitants on such a large scale, the recent coronavirus pandemic ought to be understood in the sense that the separation or supremacy of human beings over the natural world is challenged by a microscopic nonhuman agent. The breach of viral boundaries across species urges us to question the limits between the human and nonhuman realms to conclude that nonhuman "actants" may well be social entities with an agency of their own who actively participate in and shape social events. The utopian promise of the coronavirus pandemic should then be not only to promote "humility and solidarity" among human beings but also to reassess our biopolitical interactions with other nonhumans and understand our intricate interconnectedness with different entities. Embracing a more multispecies perspective and ultimately avoiding our anthropocentric mindsets would help us approach human and nonhuman others with acceptance and respect. In line with Bennett's theory, I hope this pandemic will lead us to a deeper awareness "of the complicated web of connections between bodies," urging us to treat both human and nonhuman bodies surrounding us "more carefully, strategically and ecologically" (17–18).

Notes

1 Historically, it has become common to divide pandemics into three categories: "The First Pandemic, known as the Justinianic Plague and its recurrent waves (541–c.750); the Second Pandemic, known as the Black Death (1346–53) and its recurrent waves that continued for several centuries; and the Third Pandemic that spread globally in a few years after its appearance in Hong Kong in 1894" (Varlık, "New Science" 197).

2 US President Donald Trump called the COVID-19 "Chinese virus" many times at the outbreak of the pandemic, sparking a diplomatic crisis with China. Naming the virus as "foreign" implies that the USA was facing an external threat and "uncontaminated, homogeneous and somewhat 'pure' population" had to be protected "from the filthy, malignant foreigner" (Viala-Gaudefroy and Lindaman).

3 In her talk "Plagued Legacies: Rethinking Black Death Narratives," Nükhet Varlık mentions the Orientalist and xenophobic rhetoric the plagues have long been associated with by the modern European discourse. For instance, an encyclopedia entry on plague in Paris dating 1765 suggests that "all the plagues that have appeared in Europe in the last two thousand years have been transmitted through the communications of the Saracens, Arabs, Moors or Turks with us, and none of our plagues had any other source" (qtd. in Varlık, "Plagued Legacies"). One of the first examples of dystopian fiction, Mary Shelley's *The Last Man* (1826) fictionalizes a plague which emerges from Constantinople after war and seemingly erases all human life in Europe.

4 Margaret Atwood's *Oryx and Crake* (2003) is a notable example of a dystopian novel fictionalizing a contagion. Following apocalyptic events ensuing from a global pandemic

that eradicates the human race, the protagonist, Snowman, discerns indications suggesting other humans may have survived at the end of the novel, hinting at the possibility of a new world.

Works Cited

Bennett, Jane. *Vibrant Matter: A Political Ecology of Things*. Duke UP, 2010.

Blasdel, Alex. "'A reckoning for our species': The Philosopher Prophet of the Anthropocene." *The Guardian*, 15 June 2017, https://www.theguardian.com/world/2017/jun/15/timothy-morton-anthropocene-philosopher.

Camus, Albert. *The Plague*. Translated by Robin Buss, Penguin Classics, 2013.

Defoe, Daniel. *A Journal of the Plague Year*. The Modern Library, 2001.

Gibbon, Sahra, et al. "Biosocial Medical Anthropology in the Time of Covid-19: New Challenges and Opportunities." *Medical Anthropology at UCL*, 29 April 2020, https://medanthucl.com/2020/04/29/biosocial-medical-anthropology-in-the-time-of-covid-19-new-challenges-and-opportunities/.

Goodwin, Jason. "Introduction." *A Journal of the Plague Year*, by Daniel Defoe. The Modern Library, 2001, xi–xv.

Harman, Graham. *Object-Oriented Ontology: A New Theory of Everything*. Pelican, 2018.

Judt, Robin. "Afterword." *The Plague*, by Albert Camus. Translated by Robin Buss, Penguin Classics, 2013, pp. 239–49.

Latour, Bruno. *Politics of Nature: How to Bring the Sciences into Democracy*. Translated by Catherine Porter, Harvard UP, 2004.

Lepore, Jill. "What our Contagion Fables Are Really about." *The New Yorker*, 23 March 2020, https://www.newyorker.com/magazine/2020/03/30/what-our-contagion-fables-are-really-about.

"Matching the Right." *The Economist*, vol. 445, no. 9320, 5 November 2022, p. 35. *ProQuest*, https://www.proquest.com/magazines/matching-right/docview/2731965333/se-2

McIntosh, Alastair. "Foreword." *Future Ethics: Climate Change and Apocalyptic Imagination*, Edited by Stefan Skrimshire, Continuum, 2010, pp. vii–xi.

Meis, Morgan. "Timothy Morton's Hyper-Pandemic." *The New Yorker*, 8 June 2021, https://www.newyorker.com/culture/persons-of-interest/timothy-mortons-hyper-pandemic.

Morton, Timothy. *Hyperobjects: Philosophy and Ecology after the End of the World*. U of Minnesota P, 2013.

Pamuk, Orhan. *Nights of Plague*. Translated by Ekin Oklap, Faber & Faber, 2022.

———. "What the Great Pandemic Novels Teach Us." *The New York Times*, 23 April 2020, https://www.nytimes.com/2020/04/23/opinion/sunday/coronavirus-orhan-pamuk.html.

Povinelli, Elizabeth A. "Geontologies: The Concept and Its Territories." *E-Flux*, no. 81, April 2017, https://www.e-flux.com/journal/81/123372/geontologies-the-concept-and-its-territories/.

Roberts, David. "Introduction." *A Journal of the Plague Year*, by Daniel Defoe, Oxford UP, 2010, pp. ix–xxvi.

Roy, Arundhati. "The Pandemic Is a Portal." *Financial Times*, 20 April 2020, https://www.ft.com/content/10d8f5e8-74eb-11ea-95fe-fcd274e920ca.

Sontag, Susan. *Illness as Metaphor*. Farrar, Straus and Giroux, 1978.

Varlık, Nükhet. "New Science and Old Sources: Why the Ottoman Experience of Plague Matters." *The Medieval Globe*, vol. 1, no. 1, 2015, pp. 193–227. *Project MUSE*, muse.jhu.edu/article/758497.

———. "Plagued Legacies: Rethinking Black Death Narratives." *Environment and the Middle East Series,* New York University Abu Dhabi (NYUAD) Institute, Abu Dhabi, 26 October 2022.

Viala-Gaudefroy, Jérôme, and Lindaman, Dana. "Donald Trump's 'Chinese Virus': The Politics of Naming." *The Conversation*, 13 April 2021, https://theconversation.com/donald-trumps-chinese-virus-the-politics-of-naming-136796.

"Virus." *Britannica Academic*, Encyclopædia Britannica, 18 August 2020, https://www.britannica.com/science/virus.

"Virus." *Key Contemporary Concepts,* John Lechte, Sage UK, 1st edition, 2003. *Credo Reference,* https://search.credoreference.com/articles/Qm9va0FydGljbGU6MjQ2MDM5NQ==?aid=278253.

Wood, James. "Outbreaks and Uprisings in Orhan Pamuk's 'Nights of Plague.'" *The New Yorker,* 24 October 2022, https://www.newyorker.com/magazine/2022/10/31/outbreaks-and-uprisings-in-orhan-pamuks-nights-of-plague.

World Health Organization. *WHO Coronavirus (COVID-19) Dashboard: Overview*. https://covid19.who.int/. Accessed 1 January 2023.

15 World-Building Enactments of the School Strike Movements during the Pandemic

Reading Youth Climate Crisis Movements through a Micro- and Nano-Utopian Lens

Heather McKnight

Introduction

This chapter argues that the youth climate movement, Fridays for Future (FFF), is a web of micro-utopias. In its call for action on the climate crisis and demands for entangled issues of social justice and resolutions to global inequalities, it holds creative potentiality for nano-utopian moments of spontaneous, unpredictable, non-violent self-organization (McKnight, *Chaos* 34). Such structures allowed FFF to utilize complex temporalities to sustain action at reduced levels throughout the pandemic. In FFF, resistance anticipates a solution even as it expresses despair; anxiety can garner militant optimism, and political organization becomes an open-ended utopian process (Bloch 198–99). This is demonstrable through the ambitious, creative, and increasingly complex engagements with social justice movements during the COVID-19 pandemic.

The youth climate movement combines anxiety, pessimism, horizons of hope, and disruptive action that emerges from a place of an apocalyptic now, rather than pure optimism (Friberg, *Disrupting* 9). It echoes the concrete utopianism of other radical environmental activist groups, engaging in multiple critiques and acts to disrupt the now of late-capitalism through prefigurative politics (Alberro 9). Prefiguration here means to act "as if" things were otherwise in order to move toward a transformative politics, one that may involve establishing counterhegemonic meanings and performing contentious acts (Cooper, *Towards* 909).

Planned actions of autonomous FFF groups, such as the school strikes and their pandemic alternatives, operate as micro-utopias; they are small prefigurative enactments of a different world (Cooper, *Everyday* 218). While sharing a global message, they do not aim to change things as unilaterally as European utopianism has historically (Bell 77; Becker). Darian-Smith describes micro-utopias as akin to Indigenous utopias, not assuming "singular ethical or moral visions that are applicable to everyone" (Darian-Smith 178). We see this in the variation of youth demands—for example, democratic demands for curriculum change and electoral reform are dominant in the UK and Australian movements (School Strike 4 Climate 4; *Demands*). With massive carbon footprints, the UK and USA have policy campaigns focusing on corporate legislation. In countries

DOI: 10.4324/9781003345770-20

with smaller carbon footprints, such as Nigeria, impacts of the climate crisis are already felt; heat deaths for both animals and humans lead youth campaigners to focus on eliminating heat inequality (Watts).

The varied focus under a singular theme demonstrates the youth climate movements' entangled yet diverse micro-utopian nature. Sixteen-year-old Nadia Nazar, the founder of the Zero Hour campaign, notes the different "nows" activists exist within in these micro-utopias, where "different people are impacted differently according to their identity, and… oppression like racism, patriarchy, capitalism, colonialism and ableism" noting how temporalities of urgency vary (Salter and Pressigny). The movement has, during the pandemic, proven to be self-reflexive in discourses on politics, oppression, and futurity, integrating new perspectives on diverse inequalities (Friberg, *(Con)Temporary Utopias* 8).

Underpinning these planned and entangled micro-utopian acts are brief, unexpected utopian moments that are generative of order within the FFF movement. We may regard these as nano-utopian moments. This idea builds on existing theories of process-based activist utopias as micro-utopias (Cooper, *Everyday* 31; Becker). The nano-utopian category aims to describe unplanned or spontaneous activist moments, viewing them as accelerated processes of self-organization that appear to arise out of chaotic situations or break down what we might call "far-from-equilibrium" situations (McKnight, *Chaos* 43). Under certain circumstances, "entropy itself becomes the progenitor of order"—a concept that is remarkably prescient in our times of multiple crises (Prigogine and Stengers xxi).

Defining aspects of a nano-utopian moment are threefold. Firstly, its spatiotemporality as a "singular moment" is emergent, unexpected, temporary, and accelerated. It generally occurs in far-from-equilibrium situations. Secondly, the nano-utopian moment, while disrupting one system, has within it the possibility of creating a "higher level" of order—i.e., one that reaches towards a new horizon of hope for a fairer ordering of the world for the participants. This reordering is not certain and also may not be as anticipated due to its unplanned and unexpected nature. Finally, the utopian drive of the nano-utopian moment enacts utopia as a process rather than a place; it is hopeful, open, critical, and based on material action (Bloch 5–9).

The emergent politics of a nano-utopian moment reaches for a collective, although not universalizing, idea of a better world. Relating to its final hopeful feature, it should be grounded in an ethics of non-violence and mutual recognition of each other's vulnerabilities. There must be a continuous critique of these concepts as reaching toward non-violent horizons, as the structural nature of violence is almost impossible to circumvent (Butler 5; Moylan 169). Certainly, the FFF movement is non-violent in its aims, and its accelerated nature works in temporal resistance to Rob Nixon's environmental slow violence "of delayed destruction that is dispersed across time and space, an attritional violence that is typically not viewed as violence at all" (2; FFF).

The research limitations of this chapter include dependence upon an analysis of the current research on the pandemic youth climate movement FFF, which includes a disproportionate focus on activism in the Global North and wealthy, predominately White communities (Neas et al. 7). Current research has a heavy focus on mass mobilizations and the digital. There is a tendency to disregard smaller, still impactful events and center interest on the individual activist Greta Thunberg without placing her in the historical context of youth climate activism.

Pre-pandemic Temporality—the Nano-Utopian Moment of the School Strike Initiation

Thunberg staged a solo climate protest at the age of 15 in front of the Swedish Parliament on 20 August 2018. She coined the hashtag "#FridaysForFuture" on 7 September, triggering a mass school strike movement (de Moor et al.). The origins of the current wave of youth climate movements have been located within a cascade of micro-utopian movements following this nano-utopian act. Previously, Thunberg had developed an eating disorder and stopped talking in her embodied reaction to the climate crisis; finding her voice again through this activist engagement, she moved from anxiety and anger to hope. This we may see as a process of becoming utopian, of re-learning the world to imagine how it could become radically other (Moylan 13). The resulting movement has not emerged in a social vacuum and represents continuity and change; undoubtedly, it has mobilized many young people in a way that may lead to lifelong political engagement (de Moor et al. 621).

Thunberg's nano-utopian moment is emergent from a rich history of youth climate and social justice activism. She was influenced by anti-gun protestors in Florida and by reading about Rosa Parks, a civil rights activist from the 1950s (Making Herstory). Arguably, there have been several notable micro-utopian youth climate movements; these have worked as a non-centralized web of mutually informing, entangled utopian forms. A youth climate movement was emerging in the 1980s (Waxman). Severn Cullis-Suzuki gave a speech to the UN in 1992 at the age of 12, denouncing world leaders for not tackling global warming, the content of which echoed many of Thunberg's sentiments in 2018 (Steer). It was after this that the formalization of the Climate Youth Movement saw umbrella movements forming to network the smaller, emergent groups globally. In 2015, on the first day of COP21, students worldwide joined a climate strike to demand government action, continuing a history of civil disobedience among young climate activists (de Moor et al. 622).

Emerging from this growing public pressure from youth, we can see the various ways Thunberg's protest is also nano-utopian. Its spatiotemporality as a "singular moment" is unexpected, temporary, and accelerated. In 2018, the world was dealing with a "convergence of global crises," with fuel shortages, war, many unable to access food, and climate related forced migrations and deaths (della Porta and Portos 5). Sweden was in a far-from-equilibrium state, facing wildfires and the hottest summer ever recorded (Jung et al. 1). Fires undoubtedly added to the local mobilization, the press coverage, and the hyper-accelerated response.

On 15 March 2019, it is estimated that around 1.6 million students from over 120 countries joined the school strike for climate crisis (Jung et al. 1–2).

Furthermore, in September 2019, the third FFF Global Day of Action saw 7.6 million participants go into the streets and to protest events across 185 countries; at the time, it was considered the largest globally coordinated climate protest (della Porta and Portos 1–2). FFF is now decentralized, with transnational, national, and local groups. For example, in Germany, more than 650 local groups have been established (Mucha et al. 264).

The acceleration of this movement is nano-utopian in nature but also resonates with utopic predictions of the "fast future" of activism, where "next-generation movements exploit differentials in time perception by moving ultrafast compared to the status quo" (White 188). Here we see the establishment of a "higher level" of order that reaches toward a new horizon of hope for climate action; indeed, this fits the bill of being unexpected in its scope. It created a new ordering and focus for the youth climate movement through FFF. It is collective but not universalizing, we see it emergent as a singular moment but then fracturing and generating decentralized strike action and groups.

Protest in the Darkness of the Lived Pandemic: Emergent Digital Strategies

The pandemic has added further terrifying complexity to the chaos of our far-from-equilibrium world. The escalating circumstances devastated human life and threatened social movements (Christou et al. 1–3). As lockdown kicked in, higher infection and mortality rates reduced protests on non-COVID-19 issues; links between death rates and declining protests in Europe are clearly mappable (Kriesi and Oana 15). The Austrian FFF branch demonstrated the fast-forwarding of youth, cancelling planned street protests before the government announced a lockdown, justifying this by their relationship to scientific demands for generational justice, and citing demands from experts to consider older people (Grenz and Knopp 395).

Research into pandemic mobilizations privileges the online while recognizing the interconnectedness of online and offline as hybridized or entangled. Social media has provided insights into how FFF managed to uphold activity and the ready availability for text analysis means work has taken place relatively quickly on this front (Haßler et al.; Fernandez-Zubieta et al. 4; Neas et al.). Some research has argued there was a "forced digitization" of FFF, with social media as "the only permitted site for activism" and street protests as "impossible" (Haßler et al. 2; Sorce and Dumitrica, "School Strikes" 2). As I will discuss in the following part of the chapter, the shift to an online format is not absolute.

Physical protest held a higher risk; also, it may have been seen as a move against government pandemic restrictions, thus creating a potential reputational risk for FFF. As a movement centered around in-person assembly and protest, and whose online engagement peaked during these in-person mass mobilizations, FFF was clearly at existential risk (Haßler et al. 3; Martí et al. 107). The weekly activity of FFF came to an abrupt halt alongside a "multiple crisis of protest routines," media attention shifting away from climate toward the pandemic, and lockdown preventing or limiting protest (Haßler 5). In March 2020, FFF's various media channels in different countries halted formal strikes to comply with lockdown regulations (Haßler et al. 5–6; Christou et al. 9). FFF in Finland and Romania committed to moving them online under #climatestrikeonline and the trend soon spread across the movement (Sorce and Dumitrica, "School Strikes" 6).

Focusing on FFF as an online or hybrid movement demonstrates openness and adaptability—a utopian trait that has allowed youth climate activism to prevail and emerge differently. Social media allows people ease of movement formation and action, as well as the ability to organize quickly and inexpensively in short time frames—lending itself well to the emergence of order from chaos in a nano-utopian sense (Sorce and Dumitrica, "School Strikes" 4). The connective nature of social media allows people to "perform a networked movement" without sharing a collective identity (Sorce and Dumitrica, "School Strikes" 4). The engagement of FFF with calls for public safety during the pandemic led to an emergent rather than a democratically decided decision to move to digital strikes. This singular nano-utopian moment would lead to different forms of order in the movement.

While this "forced digitisation" kept the movement alive during lockdowns, some note that this was not a source of innovative tactics (Sorce and Dumitrica, "School Strikes" 1). It echoes the process of contestation between recuperative and radical approaches emergent in many other movements during the pandemic (Firth 4). The move to digital platforms privileged community-building and education over direct action, and some worried this would soften the political impact of landmark school strikes (Sorce and Dumitrica, "School Strikes" 1). FFF activists in Germany and many other countries sustained online activity every Friday instead of street protests (Hunger; Martí et al. 108). Online engagement has its limitations; as a research participant noted, "the feeling is simply missing, that of blocking the street together with friends," producing the lack of a mass feeling of participation (Grenz and Knopp 399). Digital strikes received criticism for not involving any withdrawal process; with no labor withheld, they did not involve a "strike action," per se (Sorce and Dumitrica, "School Strikes" 7). Such criticisms are not new: school strikes have been previously represented as "truanting" or "skipping school" rather than legitimate strikes due to the media infantilization of youth movements (Hay; Burns).

The FFF school strikes included recreations of existing protest symbols, adaptation into domestic environments, and a move from public to private (Grenz and Knopp 396). Organizers encouraged participants to photograph themselves with a protest sign. They set it as their profile picture, using hashtags such as #climatestrikeonline or #digitalstrike to link with the broader social media trend (Sorce and Dumitrica, "School Strikes" 7). Online placards and digital pickets were used in many pre-pandemic campaigns; they were not revolutionary. Despite this, they created digital visibility for the movement, arguably at a time during lockdown when many more people were viewing life through their social media feeds than ever before.

Data indicates an independent decline in FFF tweets before lockdown before the pandemic. Some question whether the lockdown further accelerated a trend of declining hashtag activism around #fridaysforfuture (Haßler et al. 18). Tweets in Germany tagged #fridaysforfuture on Twitter fell by a quarter after the pandemic started (Haßler et al. 17). There was still a high volume of hashtag activism encouraging debate, establishing, and configuring "ad hoc publics" forming the foundations of campaigns to instigate policy change (Haßler et al. 7). FFF successfully managed to use new forms of online-only protests, continuing debate about climate change—albeit at a lower volume.

The fifth FFF Global Climate Strike (24 April 2020) garnered significant engagement in tweets; although they were still much lower than during similar offline protest events, they were significantly larger than during the rest of the lockdown (Haßler et al. 18; Mucha et al.; Grenz and Knopp). Nineteen thousand were present for the online live stream with contributions from activists, artists, and scientists, making this one of the most notable digital strikes ever (Hunger; Sorce and Dumitrica, "School Strikes" 4). The online modality allowed for different temporalities of engagement and, although there is little evidence this was necessarily a plus in terms of the transformative potential of the event, it did allow for demonstrable and interactive inter-protest solidarity. For example, four FFF collectives in Slovenia, Poland, Lithuania, and Hungary held synchronous events using live-streaming, an alternative digital replacement of a collective action, rather than a series of tweets or posts (Sorce and Dumitrica, "School Strikes" 7).

Arguably, the digital strike format asked supporters to perform themselves as protestors without the contentious act of actual withdrawal. This symbolic act is more easily ignored by those in positions of power; these events failed to engage as much as in-person events with political structures and therefore may lack in efficacy (Sorce and Dumitrica, "School Strikes" 12–13). Further research is needed to understand how a simple digital strike compares to the political impact of in-person mobilization. A further issue is that digital activism is embedded in corporate social media platforms and therefore at the mercy of algorithmic controls and central surveillance, which may put activists at increased risk should they not be tech savvy (Sorce and Dumitrica, "School Strikes" 4).

Online action shifted the activist focus from contention to collaboration; there was an increased focus on informational and educational resources, community-building, and partnership development, alongside further debates about the legitimacy of FFF and a decline in organizing messaging (Sorce and Dumitrica, "School Strikes" 2–3; Haßler et al.). In the Pacific Island state of Vanuatu, FFF activists relocated protests due to dual concerns of lockdown and cyclone Harold; they recorded shows for local TV and radio to air during commercial breaks (FFF). FFF online petitions and open letters were used sparingly for aviation bailouts, highlighting the importance of continuing climate policy in lockdown and local policy issues; petitions were shared from larger non-governmental organizations (NGOs) with climate agendas (Sorce and Dumitrica, "School Strikes" 8–9). FFF in North Macedonia organized four webinars and a month of educational engagements (FFF). Resistance reliant on the interruption of regular school activities was to some extent replaced, with a focus on student climate networks and building systems and strategies (Mucha et al. 271–76). Movements used this time for weekly online meet-ups and build the movement and its strategy for the future, with new narratives emerging (FFF; Murray).

#FightEveryCrisis—Responding to Chaos with Complexity

FFF North Macedonia noted that the climate crisis and the pandemic were "two topics [that] are not that different," a sentiment echoed throughout the movement (FFF). The pandemic raised public awareness of narratives at the core of FFF—worldwide interdependence and complex interconnectedness—and a further interweaving of global discourses in the movement (Grenz and Knopp 391). Online messaging from FFF identified vulnerable groups as hardest hit by both the climate crisis and the pandemic, for instance, with those in the Global South faced the harshest consequences of the climate crisis and received smaller amounts of vaccine (Wallis et al. 2).

FFF converged with other emergent nano-utopian movements, such as the pandemic mobilizations around George Floyd's murder by a police officer in Minneapolis, and the resulting decentralized micro-utopias of the Black Lives Matter (BLM) movement (Asayama et al.). In calls for environmental justice, youth movements aligned with BLM, linking colonial and capitalist oppression and violence with the climate crisis. The youth climate movements increasingly recognized the pandemic's connection with this slow violence and economic injustice of environmental destruction that worsens inequality (Nixon 70; Christou et al. 11). The movements' narratives adopted further complexification and intersectionality of issues beyond environmental protection and into political economy, a process that encourages "the transformation of both individual and social identities and socio-economic systems" (Ivanović et al. 304).

The shifting narrative of the FFF and youth climate movement led to the new "Fight Every Crisis" slogan, hashtag, and branding—a call to tackle crises simultaneously and collectively, from which a new utopian horizon emerged. Both the climate crisis and the pandemic have health, social, ecological, and economic consequences, creating moments that decide between life and death. #Fighteverycrisis depicted the urgency and entanglement of the crises (Grenz and Knopp 396; Wallis et al. 6–7). Such narratives recognize it is imperative to address socio-economic inequality for global sustainability (Asayama et al. 699). The pandemic became integrated into the FFF debate on climate crisis. It became part of their messaging in relation to social justice and the whole world life and death impact of these multiple life-threatening events (FFF; "#fighteverycrisis"; Wallis et al. 2). "[T]he non-mentioning of individual crises" re-framed normative definitions of what the movement is about: continuing support for scientific expertise and the demand for intergenerational justice to include narratives around the support of vulnerable elders (Grenz and Knopp 397). It also served to keep the movement's messaging relevant at a time when the pandemic dominated the media and climate policies were falling off the agenda.

The nano-utopian moment emerges, and new forms of order come at the moment of entanglement with other movements responding to global incidents. Linking with the issues of BLM and the pandemic were the nano-utopian turning points generated by far-from-equilibrium situations, these saw the nano-utopian emerge as well as the micro-utopian. The planned and the unplanned or emergent actions impact upon each other, and the emerging collective daydream prefigures new alternatives (Bloch 91–93). Such intersections develop the nuance of the argument, expanding the range and impact of the youth movements' language and scope.

The Creativity of Micro-utopian Resistances

While this focus on social media is significant and measurable, studies show that despite the lockdown and public health threats, there was still significant in-person collective mobilization for protest. Public action continued uninterrupted across Europe and beyond, albeit at a lower level for many movements (Kriesi and Oana 6–7; Borbáth et al. 320). Strikingly, the US mobilization of the BLM movement against racial injustice and police brutality saw between 15 and 26 million take to the streets (Kriesi and Oana 6–7; Borbáth et al. 320). In Europe, environmental activism dropped from accounting 13.7% of all European activism to 4.8% during the first wave of COVID-19 but rose to account for 9.3% in the second wave (Kriesi and Oana 6, 18). This more general rise in climate activism across Europe was attributable primarily to the FFF movement (Kriesi and Oana 20).

The breadth of available social media data and the high risk and difficulty of ethical approval for the research of in-person and physical events have produced a marginalization (and in some cases a borderline erasure) of in-person youth climate actions by researchers. Very few studies focused on overall protest levels during the crisis or the repercussions on civil society organizations (Kriesi and Oana; Mucha et al.; Borbáth et al.). One study noted that online protest and communication for FFF in Germany diminished at the same rate as the in-person events of postering, street posters, and chalk protest during lockdown (Mucha et al. 273). There seems to be little regard paid to the fact that these events are still taking place at a reduced level in far-from-equilibrium and dangerous times. The number of participants decreased for all action forms, including online petitions, and, as such, we may regard the sustainment of any in-person activity as significant (Kriesi and Oana 11).

Mapping social media posts alone fails to recognize creativity and planning the relationship between person, place, and politics that happens off and behind the screen before and around these posts. The increased reliance on digitized approaches did not "mean a disappearance of the body, but a rearrangement of the relations of bodies, objects, and places as well as their framings" (Grenz and Knopp 396). While they were no longer about the interruption of school time in the same way, youth activists found physical ways of circumventing pandemic rules for occupying space. The hologram strike in Poland saw FFF organizers project former protest images onto government buildings; and during balcony strikes in Austria, Hungary, and Bulgaria, protest signs were displayed outdoors (Sorce and Dumitrica, "School Strikes" 7). In solidarity with the FFF movement, Greenpeace activists in Warsaw staged a holographic protest; two activists wearing PPE and working at a distance of two meters from each other set up the event (Greenpeace).

FFF conducted hybrid protest actions in a digital strike for climate change on 24 April 2020, with a 24-hour live stream under the banner #FightEveryCrisis. This was to happen two days after the 50th Earth Day, which saw many other hybrid protest events (Murray). In-person activity was best documented in Germany as a stronghold in terms of both FFF activity and research and works as a useful case study (Haßler et al. 5). Two nights before the strike, in support of FFF, a Greenpeace activist released ghostly images of a holographic climate demonstration in the early morning hours in front of the Reichstag, the semi-transparent holographic projections showing scenes from the 2019 large-scale climate strike (Picco). There were 214,000 views on the German protest livestream, and over 15,000 climate placards laid in front of the parliament, from over 70 local groups all over Germany (FFF; Hunger; Haßler et al. 6). The Konstanz local group coordinated simultaneous hybrid campaigns. Banners hung from houses, local groups ran digital meetings, and they wrote 80 letters of protest to politicians stating that COVID-19 measures should still factor in environmental goals. In Kiel, a total of 140 activists held a demonstration on bicycles against a new motorway (FFF).

In Darmstadt, an action-walk took place, for which a route was published and prepared with posters by organizers through the city center; on the day, protestors could leave traces in the form of banners, posters, and chalk (FFF). In Mainz, 50 posters were sent to Berlin for the creative protest at the parliament. They also made a video for the live stream, and activists illuminated local landmarks with green light (FFF). FFF Stuttgart organized a climate walk that people could follow masked and socially distanced; in compliance with the coronavirus regulations, posters, banners, and chalk paintings were left at certain stations in the city center (FFF).

The actions were not without incident: in Darmstadt, many of the posters were torn down and, in Stuttgart, despite compliance and pre-agreed regulations being followed, participants still faced police intimidation (FFF). FFF have contested claims that offline events could not prove newsworthy during lockdown. Researchers have noted that the online/offline co-ordination can be seen as "media-savvy and transgressive" (Haßler et al. 3). A visual narrative of small and alternative resistance during lockdowns that occupy offline spaces is emergent. Offline actions on climate crisis during the pandemic, while smaller in scale, should be seen as greater in terms of the risk to individuals and the reputational risk to the movements if anything went wrong. Within these unusual circumstances, we see many small, planned activities that come into play as the youth climate movement communities pull against the current systems of order that contain uncertainty. We can read these as micro-utopias that aim not to create disarray but micro-orders within these systems, sustaining hope, generating knowledge, and building communities that may aid a further process of social change.

September 2020 saw worldwide "shoe strikes." At FFF Hyderabad, India, shoes were collected or posted to organizers and placed in spaces used for protest, thousands of participants sent old shoes with messages to the government, and afterwards, shoes were donated to NGOs (Varma). This happened in many other locations, including Tokyo, Sweden, France, the UK, the USA, and Canada ("FridaysforFuture Shoe Strike"). Perhaps even more than the placards on the ground, these walkways and town squares occupied by empty shoes, suggesting an invisible presence, carried a certain hauntology of protestors neither present nor absent in these spaces (Derrida 51).

We can view these events as micro-utopias where the shoe strikes, placard displays, holographic protests occupy spaces to explore social alternatives to doing and being through practice rather than imagination (Cooper 31). These actions cannot perhaps replace the previous FFF physical marches and occupations where "protesters take part in certain bodily forms of protest that require their sole presence" (Grenz and Knopp 394). Alternative occupations still involved a bodily performativity through which protestors visibly committed to their demands (Grenz and Knopp 395). While they have not yet formed the cornerstone of research, there is a militant optimism to these acts reaching, in their small ways, to a return to the pre-pandemic physical occupations.

Emerging Futurities

Greta Thunberg warned of the turn away from climate activism in the pandemic, which was used as a reason to pursue business as usual. This and the impossibility of returning to a form of normalcy combined with a simultaneous desire for a return to in-person marches is apparent throughout FFF as we emerge from lockdown. FFF notes how the pandemic response has proven that governments can act fast: such speedy responses could apply to climate action and have discredited ecofascist narratives emergent during the pandemic ("Demands"). Youth protests re-emphasize the immediacy of the climate crisis, adopting multifaceted creative responses. The emotions opened-up by the pandemic protest actions in FFF can have a politicizing effect, creating new forms of order, with "a stimulating potential in triggering moments of political awakening, and in opening a political imaginary made of change, bifurcation, and new types of affective identities" (Knops 207). Indeed, the FFF movement states clearly how the break from normality has fractured and opened-up new realizations: "[w]e cannot go back to normal. Normality was already a crisis—a crisis of inequality, destruction of nature, and climate. … The system is not broken; it was built to be unjust" (FFF in Friberg 8).

As a measure of the youth movement, FFF has been resilient to the pandemic situation. This has been due to its existing digital backbone and its creative adaptability; it demonstrates unifying emergence and autonomous diversity (Mucha et al. 261; Haßler et al. 18–19; Hunger). There is power in the small planned micro-utopias, and the question is: will FFF be able to continue to use this? Will this adaptability lead to continued resistance as we see increasing policing of youth-led environmental protests? (Sloam et al. 686). While writing from a place of climate privilege in the UK, we are at present gradually having our protest rights stripped from us one piece of legislation at a time; learning different modes of protest and resistance is essential. If the pandemic was a trial run for a country without free assembly rights, then perhaps there are utopian lessons to be learned across the board from FFF that will persist, ensuring we always have at least a nano-utopian move within our grasp.

Works Cited

Alberro, Heather. "In and Against Eco-Apocalypse: On the Terrestrial Ecotopianism of Radical Environmental Activists." *Utopian Studies*, vol. 32, no. 1, April 2021, pp. 36–55. *Silverchair*, https://doi.org/10.5325/utopianstudies.32.1.0036

Asayama, Shinichiro, et al. "Are We Ignoring a Black Elephant in the Anthropocene? Climate Change and Global Pandemic as the Crisis in Health and Equality." *Sustainability Science*, vol. 16, no. 2, March 2021, pp. 695–701.

Becker, Carol. *Microutopias and Pedagogies for the Twenty-First Century*. http://www.caroldbecker.com/sitepages/essays_articles/art_society/art-as-a-thinking-process.html. Accessed 28 August 2020.

Bell, David M. *Rethinking Utopia: Place, Power, Affect*. Routledge, 2017.

Bloch, Ernst. *The Principle of Hope*. Edited by Stephen Plaice and Paul Knight, vol. one, MIT Press, 1995.

Borbáth, Endre, et al. "Civic and Political Engagement during the Multifaceted COVID-19 Crisis." *Swiss Political Science Review*, vol. 27, no. 2, 2021, pp. 311–24.

Bowman, Benjamin. "Imagining Future Worlds alongside Young Climate Activists: A New Framework for Research." *Fennia—International Journal of Geography*, vol. 197, no. 2, December 2019, pp. 295–305.

Burns, Judith. *School Children "strike" over Climate Change*. 15 February 2019. *www.bbc.com*, https://www.bbc.com/news/uk-47250424. Accessed 2 July 2019.

Butler, Judith. *The Force of Nonviolence: An Ethico-Political Bind*. Verso Books, 2020.

Christou, Georgina, et al. "'The Slow Pandemic': Youth's Climate Activism and the Stakes for Youth Movements under Covid-19." *Children's Geographies*, vol. 21, no. 2 January 2022, pp. 1–14.

Cooper, Davina. *Everyday Utopias: The Conceptual Life of Promising Spaces*. Duke UP, 2013.

———. "Towards an Adventurous Institutional Politics: The Prefigurative 'as If' and the Reposing of What's Real." *The Sociological Review*, vol. 68, no. 5, September 2020, pp. 893–916. *SAGE Journals*, https://doi.org/10.1177/0038026120915148

Darian-Smith, Eve. "Decolonising Utopia." *Australian Journal of Human Rights*, vol. 22, no. 2, November 2016, pp. 167–83. *Taylor and Francis+NEJM*, https://doi.org/10.1080/1323238X.2016.11910947

de Moor, Joost, et al. "New Kids on the Block: Taking Stock of the Recent Cycle of Climate Activism." *Social Movement Studies*, vol. 20, no. 5, September 2021, pp. 619–25. *Taylor and Francis+NEJM*, https://doi.org/10.1080/14742837.2020.1836617

della Porta, Donatella. *Contentious Politics in Emergency Critical Junctures: Progressive Social Movements during the Pandemic*. Cambridge UP, 2022.

della Porta, Donatella, and Martín Portos. "Rich Kids of Europe? Social Basis and Strategic Choices in the Climate Activism of Fridays for Future." *Italian Political Science Review/Rivista Italiana Di Scienza Politica*, December 2021, pp. 1–26.

"Demands." *UKSCN*, https://ukscn.org/demands. Accessed 9 July 2019.

Derrida, Jacques. *Specters of Marx: The State of the Debt, the Work of Mourning and the New International*. Routledge, 2006.

Fernandez-Zubieta, Ana, et al. *Digital Activism Masked. The Fridays for Future Movement and the "Global Day of Climate Action."* 4200523, 12 April 2022. *Social Science Research Network*, https://dx.doi.org/10.2139/ssrn.4200523

FFF. "#FightEveryCrisis: Biggest Digital Demonstration Ever and Climate Strikes around the World." *FFF*, 25 April 2020, https://fridaysforfuture.de/2404-en/. Accessed 2 February 2023.

Firth, Rhiannon. *Disaster Anarchy: Mutual Aid and Radical Action*. Pluto Press, 2022.

Friberg, Anna. "Disrupting the Present and Opening the Future: Extinction Rebellion, Fridays for Future, and the Disruptive Utopian Method." *Utopian Studies*, vol. 33, no. 1, March 2022, pp. 1–17. *Silverchair*, https://doi.org/10.5325/utopianstudies.33.1.0001

———. "On the Need for (Con)Temporary Utopias: Temporal Reflections on the Climate Rhetoric of Environmental Youth Movements." *Time & Society*, April 2021, p. 0961463X21998845. *SAGE Journals*, https://doi.org/10.1177/0961463X21998845

"FridaysforFuture Shoe Strike." *Facebook*, September 2020, https://www.facebook.com/shoestrikeforclimatechangeAccessed 26 January 2023.

Greenpeace. *Hologram Protest Warsaw*. 2 April 2020, https://media.greenpeace.org/C.aspx. Accessed 3 February 2023.

Grenz, Tilo, and Philipp Knopp. "COVID-19, Routinedynamiken und Strukturreflexivität. Zum ereignishaften Wandel der Protestformen von Fridays for Future." Österreichische Zeitschrift für Soziologie, vol. 46, no. 4, December 2021, pp. 385–405. *Springer Link*, https://doi.org/10.1007/s11614-021-00462-z

Haßler, Jörg, et al. "Influence of the Pandemic Lockdown on Fridays for Future's Hashtag Activism." *New Media & Society*, June 2021, p. 146. *SAGE Journals*, https://doi.org/10.1177/14614448211026575

Hay, Katherine. *Brighton Pupils Skip School again for Climate Change March*. 22 June 2019, https://www.theargus.co.uk/news/17723577.brighton-pupils-skip-school-again-for-climate-change-march/. Accessed 2 July 2019.

Hunger, Sophia. "Online Strikes with the Usual Suspects: How Fridays for Future Has Coped with the Covid-19 Pandemic." *EUROPP*, 29 June 2020, https://blogs.lse.ac.uk/europpblog/2020/06/29/online-strikes-with-the-usual-suspects-how-fridays-for-future-has-coped-with-the-covid-19-pandemic/. 20 September 2020.

Ivanović, Vladimir, et al. "What Kind of Mo(ve)Ment Is Fridays for Future? Motivation, Success Perception and Climate Action Framing in Fridays for Future Croatia." *Interdisciplinary Description of Complex Systems: INDECS*, vol. 20, no. 4, August 2022, pp. 304–18.

Jung, Jieun, et al. "When a Girl Awakened the World: A User and Social Message Analysis of Greta Thunberg." *Sustainability*, vol. 12, no. 7, 7 January 2020, p. 2707.

Knops, Louise. "The Fear We Feel Everyday: Affective Temporalities in Fridays for Future." *South Atlantic Quarterly*, vol. 122, no. 1, January 2023, pp. 203–14. *Silverchair*, https://doi.org/10.1215/00382876-10242784

Kriesi, Hanspeter, and Ioana-Elena Oana. "Protest in Unlikely Times: Dynamics of Collective Mobilization in Europe during the COVID-19 Crisis." *Journal of European Public Policy*, November 2022, pp. 1–26. *Taylor and Francis+NEJM*, [DOI/link?].

Making Herstory. *Greta Thunberg: A Force for Nature*. 3 September 2019, https://makingherstory.org.uk/greta-thunberg/. Accessed 31 January 2023.

Martí, Roger Soler i, et al. "The Interdependency of Online and Offline Activism: A Case Study of Fridays for Future-Barcelona in the Context of the COVID-19 Lockdown." *Hipertext.Net*, no. 21, 21 November 2020, pp. 105–14. *raco.cat*, https://doi.org/10.31009/hipertext.net.2020.i21.09

McKnight, Heather. "Chaos and Hope: Nano-Utopian Moments of Activist Self-Organisation." *Excursions Journal*, vol. 10, no. 1, 1 December 2020, pp. 33–60.

———. "'The Oceans Are Rising and So Are We': Exploring Utopian Discourses in the School Strike for Climate Movement." *Brief Encounters*, vol. 4, no. 1, 1 March 2020. www.briefencounters-journal.co.uk.

Moylan, Tom. *Becoming Utopian: The Culture and Politics of Radical Transformation*. Bloomsbury Academic, 2020.

Mucha, Witold, et al. "Fridays for Future Meets Citizen Science. Resilience and Digital Protests in Times of Covid-19." *Voluntaris*, vol. 8, no. 2, December 2020, pp. 261–77.

Murray, Jessica. "Climate Strikes Continue Online: 'We Want to Keep the Momentum Going.'" 22 April 2020. *The Guardian*, https://www.theguardian.com/environment/2020/apr/22/climate-strikes-continue-online-we-want-to-keep-the-momentum-going. Accessed 3 February 2023.

Neas, Sally, et al. "Young People's Climate Activism: A Review of the Literature." *Frontiers in Political Science*, vol. 4, 2022. *Frontiers*, https://doi.org/10.3389/fpos.2022.940876

Nixon, Rob. *Slow Violence and the Environmentalism of the Poor*. Harvard UP, 2011.

Picco, Gail. "Painting the Light." *The Charity Report*, 29 April 2020, https://www.thecharityreport.com/photo-essay/painting-the-light/. Accessed 2 February 2023.

Prigogine, Ilya, and Isabelle Stengers. *Order out of Chaos*. Reissue, Bantam Doubleday Dell Publishing Group, 1984.

Salter, Steve, and Clementine de Pressigny. "Why Are Thousands of Students Striking from School?" *I-D*, 11 February 2019, https://i-d.vice.com/en_uk/article/kzdz7y/youth-strike-for-climate-change-greta-thunberg. Accessed 18 July 2019.

School Strike 4 Climate. "About." *School Strike 4 Climate (Australia)*, https://www.schoolstrike4climate.com/about. Accessed 18 July 2019.

Sloam, James, et al. "Young People and Environmental Activism: The Transformation of Democratic Politics." *Journal of Youth Studies*, vol. 25, no. 6, July 2022, pp. 683–91. *Taylor and Francis+NEJM*, https://doi.org/10.1080/13676261.2022.2056678

Sorce, Giuliana, and Delia Dumitrica. "#fighteverycrisis: Pandemic Shifts in Fridays for Future's Protest Communication Frames." *Environmental Communication*, vol. 17, no. 3, August 2021, pp. 1–13. *Taylor and Francis+NEJM*, https://doi.org/10.1080/17524032.2021.1948435

———. "From School Strikes to Webinars: Mapping the Forced Digitalization of Fridays for Future's Activism during the COVID-19 Pandemic." *Convergence*, December 2022, p. 135. *SAGE Journals*, https://doi.org/10.1177/13548565221148112

Steer, George. "'Climate Change Is Forever.' Thousands of British Schoolchildren Walk Out in Environmental Protest." *Time*, 15 February 2019, https://time.com/5530374/europe-climate-change/. Accessed 15 February 2015.

Thomas-Peter. "UK Carbon Emissions Target 'Doing More Harm than Good,' Teen Climate Activist Greta Thunberg Tells Sky News." *Sky News*, https://news.sky.com/story/uk-carbon-emissions-target-doing-more-harm-than-good-teen-climate-activist-greta-thunberg-tells-sky-news-11764916. Accessed 18 July 2019.

Varma, Uttara. "Unique Shoe Strike on Sept 25 to Highlight Climate Issues." *The Times of India*, 19 September 2020. https://timesofindia.indiatimes.com/city/hyderabad/unique-shoe-strike-on-sept-25-to-highlight-climate-issues/articleshow/78195295.cms. Accessed 4 February 2023.

Wallis, Hannah, et al. "#Fighteverycrisis: A Psychological Perspective on Motivators of the Support of Mitigation Measures in the Climate Crisis and the COVID-19 Pandemic." *Journal of Environmental Psychology*, vol. 84, December 2022, p. 101898. *ScienceDirect*, https://doi.org/10.1016/j.jenvp.2022.101898

Watts, Jonathan. "'The Crisis Is Already Here': Young Strikers Facing Climate Apartheid." *The Guardian*, 19 September 2019. *www.theguardian.com*, https://www.theguardian.com/environment/2019/sep/19/the-crisis-is-already-here-young-strikers-facing-climate-apartheid. Accessed 20 September 2020.

Waxman, Oliva. "What Happens When Young Environmental Activists Grow Up." *Time*, 15 March 2019, https://time.com/5549331/youth-climate-strike-ecokids/. 2 February 2023.

White, Micah. *The End of Protest: A New Playbook for Revolution*. Knopf Canada, 2016.

Index

Note: Page numbers followed by "n" refer to end notes.

For Product Safety Concerns and Information please contact our EU
representative GPSR@taylorandfrancis.com
Taylor & Francis Verlag GmbH, Kaufingerstraße 24, 80331 München, Germany

www.ingramcontent.com/pod-product-compliance
Lightning Source LLC
LaVergne TN
LVHW021123300926
849160LV00010B/845
* 9 7 8 1 0 3 2 3 8 5 9 3 8 *